ASCE Manuals and Reports on Engineering Practice No. 116

Navigation Engineering Practice and Ethical Standards

A Task Committee of the
Waterways and Navigation Engineering
Committees of the Coasts, Oceans, Ports, and Rivers Institute
of the American Society of Civil Engineers

Edited by
William H. McAnally, Ph.D., P.E.

ASCE American Society of Civil Engineers

Library of Congress Cataloging-in-Publication Data

Navigation engineering practice and ethical standards / edited by
William H. McAnally.
p. cm. — (ASCE manuals and reports on engineering practice ;
no. 116)
Includes bibliographical references and index.
ISBN 978-0-7844-0992-3
1. Hydraulic engineering—Moral and ethical aspects. 2. United
States. Army. Corps of Engineers. 3. Hydraulic engineering—United
States. 4. Channels (Hydraulic engineering) 5. Hydraulic structures.
6. Engineering design—United States. I. McAnally, William H.
TC423.N355 2008
627.0973—c22

2008036680

Published by American Society of Civil Engineers
1801 Alexander Bell Drive
Reston, Virginia 20191

www.pubs.asce.org

ISBN 13: 978-0-7844-0992-3
Manufactured in the United States of America.

16 15 14 13 12 11 10 09 1 2 3 4 5

MANUALS AND REPORTS ON ENGINEERING PRACTICE

(As developed by the ASCE Technical Procedures Committee, July 1930, and revised March 1935, February 1962, and April 1982)

A manual or report in this series consists of an orderly presentation of facts on a particular subject, supplemented by an analysis of limitations and applications of these facts. It contains information useful to the average engineer in his or her everyday work, rather than findings that may be useful only occasionally or rarely. It is not in any sense a "standard," however; nor is it so elementary or so conclusive as to provide a "rule of thumb" for nonengineers.

Furthermore, material in this series, in distinction from a paper (which expresses only one person's observations or opinions), is the work of a committee or group selected to assemble and express information on a specific topic. As often as practicable, the committee is under the direction of one or more of the Technical Divisions and Councils, and the product evolved has been subjected to review by the Executive Committee of the Division or Council. As a step in the process of this review, proposed manuscripts are often brought before the members of the Technical Divisions and Councils for comment, which may serve as the basis for improvement. When published, each work shows the names of the committees by which it was compiled and indicates clearly the several processes through which it has passed in review, in order that its merit may be definitely understood.

In February 1962 (and revised in April 1982) the Board of Direction voted to establish a series entitled "Manuals and Reports on Engineering Practice," to include the Manuals published and authorized to date, future Manuals of Professional Practice, and Reports on Engineering Practice. All such Manual or Report material of the Society would have been refereed in a manner approved by the Board Committee on Publications and would be bound, with applicable discussion, in books similar to past Manuals. Numbering would be consecutive and would be a continuation of present Manual numbers. In some cases of reports of joint committees, bypassing of Journal publications may be authorized.

MANUALS AND REPORTS ON ENGINEERING PRACTICE CURRENTLY AVAILABLE

No.	Title
40	Ground Water Management
45	Consulting Engineering: A Guide for the Engagement of Engineering Services
49	Urban Planning Guide
50	Planning and Design Guidelines for Small Craft Harbors
54	Sedimentation Engineering
57	Management, Operation and Maintenance of Irrigation and Drainage Systems
60	Gravity Sanitary Sewer Design and Construction, Second Edition
62	Existing Sewer Evaluation and Rehabilitation
66	Structural Plastics Selection Manual
67	Wind Tunnel Studies of Buildings and Structures
68	Aeration: A Wastewater Treatment Process
71	Agricultural Salinity Assessment and Management
73	Quality in the Constructed Project: A Guide for Owners, Designers, and Constructors
77	Design and Construction of Urban Stormwater Management Systems
80	Ship Channel Design
81	Guidelines for Cloud Seeding to Augment Precipitation
82	Odor Control in Wastewater Treatment Plants
84	Mechanical Connections in Wood Structures
85	Quality of Ground Water
91	Design of Guyed Electrical Transmission Structures
93	Crane Safety on Construction Sites
94	Inland Navigation: Locks, Dams, and Channels
95	Urban Subsurface Drainage
97	Hydraulic Modeling: Concepts and Practice
98	Conveyance of Residuals from Water and Wastewater Treatment
100	Groundwater Contamination by Organic Pollutants: Analysis and Remediation
101	Underwater Investigations
103	Guide to Hiring and Retaining Great Civil Engineers
104	Recommended Practice for Fiber-Reinforced Polymer Products for Overhead Utility Line Structures
105	Animal Waste Containment in Lagoons
106	Horizontal Auger Boring Projects
107	Ship Channel Design (Second Edition)
108	Pipeline Design for Installation by Horizontal Directional Drilling
109	Biological Nutrient Removal (BNR) Operation in Wastewater Treatment Plants
110	Sedimentation Engineering: Processes, Measurements, Modeling, and Practice
111	Reliability-Based Design of Utility Pole Structures
112	Pipe Bursting Projects
113	Substation Structure Design Guide
114	Performance-Based Design of Structural Steel for Fire Conditions
115	Pipe Ramming Projects
116	Navigation Engineering Practice and Ethical Standards

CONTENTS

PREFACE

This manual was produced by a Task Committee of the Waterways Committee and Navigation Engineering Committee of the Coasts, Oceans, Ports, and Rivers Institute, American Society of Civil Engineers.

Members of the Task Committee and authors of this manual are:

- William H. McAnally, Ph.D., P.E., F.ASCE, Associate Professor of Civil and Environmental Engineering at Mississippi State University, teaches and performs research in navigation engineering. Primary author of Chapter 7 and editor of this volume.
- Bruce L. McCartney, P.E., M.ASCE, U.S. Army Corps of Engineers, retired. Committee Chairman and primary author of Chapters 3, 4, 5, 6, 9, and 10.
- Charles C. Calhoun, Jr., P.E., F.ASCE, Consultant, Retired Deputy Director of the Corps's Coastal and Hydraulics Laboratory, conducts leadership and ethics programs for private and public sector organizations, including ASCE. Primary author of Chapter 2.
- Michael D. Cox, M.COPRI, Chief, Lock and Dam Section, Illinois Waterway Project Office, U.S. Army Corps of Engineers, Rock Island District. Primary author of Chapter 8.
- Thomas J. Pokrefke, MSCE, P.E., M.ASCE, Consultant, Hydraulic Engineer. Primary author of Chapter 11.

The authors express their appreciation to the reviewers of this Manual, who included: Thomas W. Wakeman, Center for Maritime Systems, Stevens Institute of Technology; Nicholas Pansic, MWH Natural Resources, Industry & Infrastructure; Dennis O. Norris, U.S. Army Corps of Engineers; Gregory Johnson, Bergmann Associates; Robert Engler, Moffat & Nichol; Eric Christensen, U.S. Coast Guard; Mark Lindgren, U.S. Army Corps of Engineers; Jim Blanchar, U.S. Army Corps of Engineers; Stephen A. Curtis, Tetra Tech EC, Inc.; Billy Edge, Texas A & M University; Michael F. Garrett, URS Corporation; David P. Devine, University of Notre Dame; Norma

Jean Mattei, University of New Orleans; and Ryan C. Murphy, Michels Corporation.

We thank the U.S. Army Corps of Engineers, the U.S. Coast Guard, and the National Oceanic and Atmospheric Administration for use of their excellent publications and graphics, which are used extensively in this manual, and Carol A. McAnally for her editorial reviews.

CHAPTER 1
INTRODUCTION

1.1 BACKGROUND AND PURPOSE

1.1.1 Background

By Congressional decree, the U.S. Army Corps of Engineers was given authority and funds to build and maintain inland waterways for navigation, ship channels for ocean-going vessels, and numerous small boat harbors. In the past, ethical considerations for navigation project design criteria were self-contained knowledge within the Corps design community. The design philosophy and design criteria were often verbally passed from senior engineers to junior engineers. The few criteria that existed were scattered throughout several Engineer Manuals and Regulations.

ASCE Manual No. 50, Report on Small Craft Harbors (1969), was the first attempt to consolidate some of the Corps's navigation criteria for small boat harbors. Consolidation of criteria for inland barge navigation systems and deep-draft ship channels was undertaken by the Corps in the 1970s. This effort resulted in the publication of Layout and Design of Shallow Draft Waterways, EM 1110-2-1611 (1980), and Hydraulic Design of Deep Draft Navigation Projects, EM 1110-2-1613 (1983).

Until recently, the Corps was the exclusive designer and maintainer of navigation channels in the United States. However, with the current move to contract out design and privatize many government missions, there has emerged a private sector audience that can benefit from past experience and lessons learned.

Unfortunately, in the 1980s and 1990s there was a government-wide initiative to reduce federal regulations. The Corps manuals were vulnerable to this purge. The ASCE Waterways Committee was aware of the potential loss of this valuable design information and undertook a preservation mission.

This Ethics Manual, along with ASCE Manuals No. 94, Inland Navigation: Locks, Dams, and Channels (McCartney et al. 1995), and No. 107, Ship Channel Design and Operation (McCartney et al. 2005), presents not only Corps navigation design practice and experience, but also foreign country practice and activities of other U.S. agencies with navigation missions.

This Manual is intended to be a reference to explain the ethical roots of navigation engineering criteria. The target audience includes beginning engineers in the Corps, private sector engineers in the United States and overseas, other U.S. government agencies involved with navigation, and university students pursuing navigation-related studies.

1.1.2 Purpose

The purpose of this Manual is to present engineering criteria and practices for design, operation, and management of navigation projects, and demonstrate how those criteria and practices are interwoven with engineering ethics.

The levee failures during Hurricane Katrina (2005) raised many questions, including engineering criteria suitability, level of protection decisions, and risk assessment. Although levees are generally considered an element of a flood control project, the same questions arise in navigation projects. These Katrina-related questions point out the need to explain the origin of design, the design process to consider criteria and risk, and project operation needed to achieve the design goals. They also clearly point to a need for ethical decision-making at every level. During the design process, pressures to reduce cost can threaten safety, efficiency, and reliability. This Manual supports adherence to sound criteria by showing how engineering ethics is interwoven into navigation project design and operation to achieve the objective of public safety. This Manual differs somewhat from the usual "how-to-do-it" format by including a "why-we-do-it" aspect, which includes an historic perspective on criteria development.

1.2 NAVIGATION PROJECTS

Navigation projects provide for waterborne transport of people and goods—by ships, barges, ferries, and other vessels. They consist of ports, harbors, channels, locks, and related facilities, and they constitute vital links in the U.S. Marine Transportation System—a collection of people, facilities, organizations, and equipment that work together to move people and goods from origin to destination using waterborne carriers for at least one component of the journey.

Navigation projects include channels for ships, barges, and other watercraft. For the purposes of this Manual, they also include the water side of ports and small boat harbors such as marinas and fishing harbors. Ship channels in the United States pass about 25 billion tons of international maritime trade cargo each year. They also provide a vital link between the home ports of our Army, Navy, and Coast Guard fleets and the oceans.

The U.S. inland waterway system consists of more than 25,000 miles (40,235 kilometers)[1] of channels and 220 lock and dam projects. This system moves commodities to and from coastal ports and interior cities. There are more than 400 small boat harbors designed and built by the Corps in coastal areas, rivers, and the Great Lakes. These harbors shelter both recreational craft and commercial fishing boats. Thousands of terminals and marinas are operated by local governments and private concerns.

The key federal players in the U.S. Marine Transportation System operations are:

- U.S. Army Corps of Engineers (USACE)
  Channel maintenance (dredging and dredging reduction), lock and dam operation and maintenance, and jetty repair.
- U.S. Coast Guard (USCG)
  Safety issues, aids to navigation, accident assessment, icebreaking, search and rescue, and vessel traffic control at some locations.
- National Oceanic and Atmospheric Administration (NOAA)
  Coastal charting, Global Positioning System, and weather forecasts.

Navigation projects are engineered—designed, constructed, operated, and maintained—in accordance with engineering criteria defined in laws, regulations, codes, guidance, and good practice. The common foundation that supports these criteria is engineering ethics.

1.3 NAVIGATION ENGINEERING

Navigation engineering involves the planning, design, construction, operation, and maintenance of safe, reliable, efficient, and environmentally sustainable navigable waterways (channels, structures, and support systems) used to move people and goods by waterborne vessels.[2]

This Manual joins three other ASCE navigation manuals to form the beginnings of a body of technical literature for development of a navigation engineering specialty in the civil engineering profession. The other

---

[1]About 12,000 miles (19,300 kilometers) of the total are federal channels.
[2]Adopted by the Navigation Engineering Subcommittee of the Coasts, Oceans, Ports, and Rivers Institute's Waterways Committee.

three manuals in the ASCE series of Manuals and Reports on Engineering Practice are:

- No. 50, Planning and Design Guidelines for Small Craft Harbors (ASCE 1994);
- No. 94, Inland Navigation: Locks, Dams, and Channels (McCartney et al. 1995); and
- No. 107, Ship Channel Design and Operation (McCartney et al. 2005).

The goal of a navigation engineering specialty is to train and support the engineers who will be designing and operating the nation's waterways in the future.

1.4 ORGANIZATION OF THE MANUAL

This Manual is organized into 12 chapters. Chapter 2 presents ethical concepts as expressed in codes of ethics adopted by many organizations. Chapters 3 through 6 examine design criteria and the design process and how they are rooted in engineering ethics. Chapter 7 describes sustainable development and how it has evolved from an ethical requirement to obey the law to a proactive stewardship ethic reflected in codes. Chapters 8, 9, and 10 present the roles of three organizations—the U.S. Army Corps of Engineers, the U.S. Coast Guard, and the National Oceanic and Atmospheric Administration, respectively—in managing the nation's navigable waterways. Chapter 11 examines the tools available for engineering analyses and how ethical requirements affect use of those tools. Chapter 12 offers concluding remarks.

CHAPTER 2
ENGINEERING ETHICS

2.1 GENERAL

Engineering projects have a profound impact on our lives. The American Society of Civil Engineers (ASCE) founders recognized this when the Society was established in 1852. At that time, civil engineering simply distinguished itself from military engineering and, consequently, all engineering disciplines were represented by the Society (the original name of ASCE was the American Society of Civil Engineers and Architects). The ASCE came to be long before there were registration laws, and there were few engineering schools. Most engineers learned the profession from other engineers and could call themselves engineers only after they had established their credentials for technical competence and integrity. ASCE membership was to be ". . . adequate proof an individual is qualified to practice" and integrity was "ensured through judicious membership requirements." ASCE membership was granted only to "the most respectable and respected members of their calling" and the founding fathers personally evaluated individuals and decided on admittance. Pfatteicher (2003) presents the history of the development of the ASCE Code of Ethics.

2.2 ENGINEERING CODES OF ETHICS

Although there have been enormous changes since its creation, ASCE is still committed to achieving the highest levels of ethical conduct. The Society recognizes that engineers have always relied on not only their own reputations, but also on the reputation of the profession as a whole. Engineers are members of a "learned profession" and, as such, are "permitted" by the public (via government-issued license) to make decisions based upon their education, experience, and ultimately upon the confidence the public has in the profession. Public confidence cannot be overemphasized. Consequently, to preserve the high ethical standards (competence and integrity) of the civil engineering profession, the Society and other professional organizations maintain and enforce strict codes of ethics. State licensing boards, authorized by law to regulate the practice of engineering, have adopted mandatory codes of ethics that usually parallel the ASCE Code of Ethics (henceforth, "the Code"). For example, the Texas Board of Professional Engineers Code of Ethics begins with, "Engineers shall be entrusted to protect the health, safety, property, and welfare of the public in the practice of their profession" (Texas 2006).

The ASCE Code gives the rules and standards members are expected to abide by. Members must subscribe to the Code and it is their duty to report promptly to ASCE any observed violation of the Code in their own practice or the practice of another. The entire Code (given in the Appendix and can also be downloaded from the ASCE web page at http://www.asce.org/inside/codeofethics.cfm) is composed of three distinct parts:

- Fundamental Principles—The principles guiding the entire Code.
- Fundamental Canons—The seven rules or "laws" that must be followed.
- Guidelines to Practice—Further guidance on each of the Canons.

Selected Canons from the Code are discussed below.

Canon 1

The Code [as well as the National Society of Professional Engineers (NSPE) Code] is emphatic about the engineer's first and primary duty when Canon 1 states,

> Engineers shall hold paramount the safety, health, and welfare of the public . . . in the performance of their professional duties.

The word "paramount" was carefully chosen for the ASCE and NSPE Codes to make absolutely clear that this duty transcends all others. The implications of this duty are immense with regard to waterways projects. Most of these projects are enormous in scope, and funding and failures of the project to perform as designed can have cataclysmic impacts on life, property, and the environment. President Herbert Hoover (1929–1933), who himself was an engineer, addressed this point when he said:

> The great liability of the engineer compared to men of other professions is that his works are out in the open where all can see them. His acts, step-by-step, are in hard substance. He cannot bury his mistakes in the

> grave like the doctors. He cannot argue them into thin air and blame the judge like the lawyers. He cannot, like the architects, cover his failures with trees and vines. He cannot, like the politicians, screen his shortcomings by blaming his opponents and hope the people will forget. The engineer simply cannot deny he did it. If his works do not work, he is damned. . . . To the engineer falls the job of clothing the bare bones of science with life, comfort, and hope. (Walesh 2000)

It is incumbent on engineers and on members of the project teams to have the technical competence to, as required, plan, design, construct, operate, and maintain waterways systems and the integrity to ensure the "safety, health, and welfare of the public." They must also work to educate clients and the public as to the risk of failure and the consequences of inaction. The current Code speaks to the required factors.

Our commitment to the environment is also recognized in Canon 1 with the requirement that engineers "shall strive to comply with the principles of sustainable development. . . ." The Code defines sustainable development as the "challenge of meeting human needs for material resources, industrial products, energy, food, transportation . . . while conserving and protecting environmental quality and the material resource base essential for future development" (See Chapter 7).

Canon 2

Canon 2 requires that "Engineers shall perform services only in areas of their competence." Guideline 2a to this Canon explains that the word "competence" refers to both education and experience. In light of rapidly changing technology, competence becomes a function of time. Components of our knowledge base can become obsolete and some segments of our experience base can become almost irrelevant.

Canon 4

Canon 4 states, "Engineers shall act in professional matters for each employer or client as faithful agents or trustees, and shall avoid conflicts of interest." It presents a potential ethical dilemma when the employer/client's interests are in conflict with public interest (e.g., if cost-cutting to make a project economically viable makes it unsafe). In such cases and in all others, the requirement of Canon 1 to hold public safety and welfare paramount must be the deciding factor, even if the client believes it not to be in his interest.

Canon 6

The public must trust that engineers are technically competent and have the integrity and character to properly apply the latest, appropriate technology to these projects that impact their lives. As previously noted, the engineer's personal reputation is closely dependent on the reputation of the profession and vice versa. A person of the highest ethical character will be tainted if working within known unethical surroundings. One unethical individual in an otherwise ethical organization can be the prism through which the entire organization is viewed. Canon 6 makes it clear that engineers "shall act in such a manner as to uphold and enhance the honor, integrity, and dignity of the engineering profession. . . ." In 2006, the following was added, ". . . and shall act with zero tolerance for bribery, fraud, and corruption."

Canon 7

In order to meet the competence requirement in Canon 2, Canon 7, which says, "Engineers shall continue their professional development . . . and shall provide professional development for engineers under their supervision" must be met. Guideline 7a further advises that "Engineers should keep current in their specialty fields by engaging in professional practice, reading in the technical literature and attending professional meetings and seminars." In addition to Canon 7, many states and organizations have specific requirements for continuing education. The bottom line is that technical competence is an absolute necessity to "hold paramount the safety, health, and welfare of the public" and is a continuing process.

As stated previously in this section, professional organizations other than ASCE maintain and enforce strict codes of ethics. For example, the NSPE code of ethics may be found at http://www.nspe.org/ethics/codeofethics/index.html.

2.3 ETHICS AND ENGINEERING PRACTICE

Guidance is readily available from ASCE, NSPE, other professional organizations, state license boards, etc. to resolve ethical situations or dilemmas the engineer may face. ASCE has an Ethics Hotline (800-548-ASCE, ext. 6061) for help. There is also guidance on the ASCE web page, including the document "Ethics—Guidelines for Conduct for Civil Engineers" (ASCE 2008).

An ethical dilemma can be defined as a situation where two or more moral ideas come into conflict and it appears all of them cannot be respected. But how often does this really happen? How many times do we know deep down what we should do—the right thing—and are really trying to justify what we want to do? The well-known ethicist Michael Josephenson said, "[E]thics is all about how we meet the challenge of doing the right thing when that act will cost more than we want to pay"

(Maxwell 2003). There is no doubt that true ethical dilemmas exist in the engineering profession. However, the view that such dilemmas are rare and the individual knows the right thing to do is widely held (see, for example, Drucker 1999, Covey 1998, Maxwell 2003, Goldberg 2006, Huntsman 2005, Garrett 2006, Veach 2006, Welch[1] 2003, and Calhoun 2007).

A legal act is not necessarily ethical. The Russian dissident Alexandr Solzhenitsyn said, "I have lived my life in a society where there was no rule of law. And that is a terrible existence. But a society where the rule of law is the only standard of ethical behavior is equally bad." Billionaire Jon Huntsman, the founder of the world's largest privately owned chemical company, said, "We are not always required by law to do what is right and proper. Decency, for instance, carries no legal mandate" (Huntsman 2005).

In this modern society there is no argument that both the law and ethics are essential to the engineering profession. Practicing engineer Michael Garrett clearly separated the two and shows their philosophical differences:

- Code of Ethics—A code of conduct to which a person voluntarily adheres because it reflects his or her values and is believed to be beneficial to society. A guide that, with rare exceptions, does not provide specific instruction. You will ask yourself—Is it the right thing to do?
- Law—Laws are intended to achieve specific behaviors without the application of either judgment or conscience. They provide specific instructions with specific punishments. You may ask yourself—What is the minimum I am required to do, or even, what is the chance I will get caught? (Adapted from Garrett 2006)

Basically, if the act is against the law it is practically always unethical, but just being legal does not mean it is the ethical or right thing to do. Legal is not synonymous with ethical in the engineering community. In the case where an engineer believes that laws are being broken and his or her advice to the contrary is being ignored, he or she has the responsibility to report the situation to the appropriate authorities under "whistleblower" rules (ASCE 2008).

The Code requires you to "do the right thing" and there may be a tendency to try to rationalize what is right versus simply what you want to do. This tendency can be enhanced when a code becomes too wordy or complex. The code should be considered a "Thou shall" rather than a "Thou shall not" document. Goldberg warns that ethics "is the study of

[1]Welch expressed these ideas in the cited book but also more forcefully in his 2001 Dolan Lecture at Fairfield University.

right and wrong . . . and twists and turns of sophisticated intellectual debate obscure a much simpler truth." Veach asks, "Have we (engineers) . . . made the concept of ethical behavior so complex and confusing that we fail to act in ways consistent with moral principles . . .?" Huntsman says that ethics are not rocket science, they are "child's play." Covey sums it up very well when he says, before you make a decision ask "[I]s it right?"

This document provides technical guidance on the planning, design, construction, operation, and maintenance of navigation projects that meet the Canon 1 requirements to hold paramount the safety, health, and welfare of the public while striving to comply with the principles of possessing a high degree of competence. Since ethics include both competence and character components, these guidelines must be applied with integrity—ensuring that sound and appropriate engineering principles are applied, thorough review procedures are carried out, and proper controls are in place and enforced.

2.4 OTHER CODES OF ETHICS

Virtually every profession professes a code of ethics [for example, the Center for Study of Ethics in the Professions (2007) lists hundred of codes] and most of them contain elements very similar to those of the engineering profession, allowing for differences in disciplines. For example, the Building Owners and Managers Institute International Code of Ethics has as its Article 1:

> Each designee of the Institute shall conduct business in a manner displaying the highest degree of professional behavior, bringing credit to the profession, the industry, and the Institute. Designees shall speak truthfully and act in accordance with accepted principles of honesty and integrity. A designee shall endeavor to understand and fairly represent his or her own scope of knowledge and ability to perform services. (BOMI 2007)

Even for individuals without a professional connection, few would argue that these ethical precepts are unwise or unjust. Even if we have not agreed to be bound by them, if we knowingly choose to violate them our only defense will be in a legalistic claim of a lack of criminal culpability.

CHAPTER 3
DESIGN PHILOSOPHY AND GOALS

3.1 GENERAL

Engineering design philosophy for navigation projects and the individual design goal for a specific project further refine the call for ethical behavior stated under Canon 1 of the ASCE Code of Ethics concerning "the safety, health, and welfare of the public." Engineering criteria can be expressed in terms of general principles, such as safety, effectiveness, and efficiency, and in more specific terms such as a 1 in 100 probability of accidental property damage, or a benefit-to-cost ratio less than 1.0. Safety and public welfare are cited as basic canons in codes of ethics, whereas other measures such as benefit-to-cost ratio are defined by law, regulations, and guidance. Good engineering practice, including proper use of mathematical and software tools, is defined by the evolving knowledge and skills of the engineering community, in which engineers are ethically required to maintain currency as described in Chapter 2.

Engineering criteria are frequently expressed in terms of these principles:

- Safe—Does not cause accidental damage to people or property.
- Effective—Achieves the design objectives fully.
- Efficient—Minimizes expenditure of resources versus benefits.
- Cost-effective—First cost and annual expenses are either commensurate with the goals or provide an acceptable rate of return on investment.
- Reliable—Minimizes downtime due to malfunction, damage, or environmental conditions.
- Robust—Handles design events that exceed design conditions without catastrophic failure.
- Environmental Sustainability—Maintains the quality and availability of natural resources for future generations.

- Social Value—Achieves societal goals.
- Aesthetic Value—Appeals to the senses.
- Secure—Prevents damage from vandalism or terrorism.

Of these principles, safety and environmental sustainability are clearly required by Canon 1 of the Code. The rest, save perhaps aesthetics, fall primarily under Canon 4 of the Code, requiring faithful stewardship of the client's interests, whether that client is the public or a private entity. Aesthetic value is the only item in this list not directly related to ethical considerations but it may be imposed by community standards. For example, the aesthetic value of riprap-protected shorelines poses an opportunity for vigorous debate that may test the limits of civil behavior, if not ethical behavior.

Another principle sometimes invoked is resilience, which combines aspects of reliability and robustness and is intended to provide for rapid recovery after a catastrophe.

3.2 DESIGNING FOR SAFETY

The design philosophy of public safety has always been used for navigation project design and has been implied, but not specifically stated, in many published works. The reason for the lack of need of formal statements is that the U.S. Army Corps of Engineers has been the principal designer and operator of the U.S. inland navigation system and coastal navigation channels for more than 100 years.

In the past, the Corps's in-house design and operations activities passed the public safety design philosophy down from senior engineers to beginning engineers. This system worked very well until the 1960s and 1970s, when there was a push to contract out a large portion of the design work (outsourcing) and reduce internal reviews. This process interrupted the informal design philosophy communications channel, and the need for specifically stated design philosophy in Corps design publications became evident.

3.3 CORPS OF ENGINEERS DESIGN GUIDANCE

USACE Engineer Regulations (ERs) outline what needs to be done and Engineer Manuals (EMs) show how to do it. The first brief USACE reference to the public safety philosophy was contained in Engineer Manual (EM) 1110-2-1611, Layout and Design of Shallow-Draft Waterways (USACE 1980), paragraph 2-11 as follows:

> Economy should consider both first cost and maintenance and operation cost without sacrifice of safety, efficiency, and dependability.

This brief statement on the importance of public safety was expanded by paragraph 4 of regulation ER 1110-2-1404, Hydraulic Design for Deep Draft Navigation Projects (USACE 1981)

> Design Rationale: The design of a deep draft navigation project must result in a safe, efficient, reliable and least cost plan with appropriate consideration of environmental and social aspects. However, the factors of safety, efficiency and reliability must be accommodated before the cost is optimized. Costs include construction, maintenance and replacement.

This first published account of the "safety, efficiency and reliability" philosophy is no longer in print but has been replaced by a revised ER 1110-2-1404, (USACE 1996c). This revised ER carries forward the same concept as follows in Section 5, Project Rationale:

> The design of a deep-draft navigation project must result in a plan that provides for a safe, efficient, reliable, and economically justified project with appropriate consideration of environmental and social aspects.
>
> a. Safety concerns the potential hazard to life and property, resulting from the consequences of ship to ship, ship to bridge, ship to moorage, and moored vessel interactions, etc.
> b. Efficiency is the optimal combination of channel, turning basin, and anchorage depths, widths, and alignments to allow traverses and maneuvers at normal speeds considering weather, waves, currents, and traffic congestion with minimal assistance for support vessels.
> c. Reliability involves the ability to achieve project purposes and proper functioning of facilities such as aids to navigation, bridge pier fendering, jetties, dikes, breakwaters, etc.
> d. Economic justification is based on the initial operational maintenance, repair, rehabilitation, and replacement costs optimized on an annual cost basis.
> e. Environmental and social aspects comprise fish-and-wildlife protection and restoration, recreational opportunity development, water quality restoration, human resources protection, and wetland preservation and mitigation of adverse aspects, etc.

Other currently available Corps publications related to navigation projects that repeat the public safety concept include:

- ER 1110-2-1457, Hydraulic Design of Small Boat Navigation Projects (USACE 1985)
- ER 1110-2-1407, Hydraulic Design for Coastal Shore Protection Projects (USACE 1997)

- ER 1110-2-1458, Hydraulic Design of Shallow Draft Navigation Projects (USACE 1998)
- EM 1110-2-1613, Hydraulic Design Guidance for Deep-Draft Navigation Projects (USACE 2006a)
- EM 1110-2-2602, Planning and Design of Navigation Locks (USACE 1995a).

These publications are available from the Corps on-line (portable document format, PDF) or printed copies. Instructions for obtaining these publications are shown on the Corps Web site, www.usace.army.mil/usace-docs/ under "Publications of the Headquarters, United States Army Corps of Engineers."

3.4 ASCE DESIGN GUIDANCE

A series of ASCE publications has been generated to convey the philosophy of public safety to the larger national and international engineering community. These publications include the following ASCE Manuals:

- No. 80, Report on Ship Channel Design (McCartney et al. 1993);
- No. 94, Inland Navigation: Locks, Dams and Channels (McCartney et al. 1995); and
- No. 107, Ship Channel Design and Operation (McCartney et al. 2005) (an expanded revision of Manual No. 80).

3.5 PIANC DESIGN GUIDANCE

The International Navigation Association (PIANC[1]) addresses the issue of safety for navigation project design in their June, 1997 publication, "Approach Channels, A Guide for Design" (PIANC 1997). It provides the following in Section 2.3:

> It is implicit in this process that the Concept Design Method should provide adequate navigational safety in accordance with good modern practice. It contains within it the implied safety margins used in many ports throughout the world.
>
> A thorough analysis of ship accidents shows that only a small percentage of accidents and marine casualties in approach channels and ports are due to channel design, but it is essential, with future commercial, economic and environmental pressures placed on port operators, that this percentage remains low.

---

[1]The International Navigation Association was formerly known as the Permanent International Association of Navigation Congresses and retains the acronym PIANC.

3.6 ETHICAL CONSIDERATIONS

The ethical mandate to place public safety and welfare as a paramount concern is obviously supported by the design criteria described here. Less obvious, but still required, are mandates for the other design principles listed in Section 3.1, such as effectiveness and cost-effectiveness, which satisfy ASCE Canon 4 to act as a faithful agent of the client; and achieving societal goals, which are expressed through laws. These ethical requirements create the potential for ethical dilemmas when public and private interests are in opposition. For example, balancing project cost versus public safety or project effectiveness versus environmental sustainability may become a matter of judgment. In such cases both individual decisions and organizational policy must discern the ethical requirements and act on them.

CHAPTER 4
DESIGN CONDITIONS AND ASSUMPTIONS

4.1 DESIGN CONDITIONS

A navigation project design needs a set of hydraulic and weather conditions before design of project features can proceed. These "design conditions" include water levels, wind velocity, wave height, and current velocity, all of which affect navigability. Design conditions have a frequency of occurrence. For example, a 100-year flood (probability of occurring in any one year equals 0.01) will produce a specific water level at the project site and a 500-year flood will produce a higher water level. Normally it is not technically or economically feasible to build a project that would withstand a flood or other event (hurricane, earthquake, etc.) of biblical proportions, so a lesser, but infrequent, event is selected for the project design event. For example, project features which affect human safety are usually designed for a 100-year event or even the Probable Maximum Flood, whereas features that affect only convenience or small property damage may be designed for a 25-year event.

Some examples for design conditions presented in Corps publications follow. ER 1110-2-1404, Deep-Draft Navigation Projects Design (USACE 1981), states:

> Design Conditions. The factors evaluated in the following paragraph provide a basis for selecting the design conditions for each layout. These design conditions must reflect weather and hydraulic conditions which are infrequently exceeded during the navigation season. The channel design usually allows for safe passage of the design vessel under most weather conditions with a competent pilot or captain. Extreme weather and hydraulic conditions are analyzed for effect on the design vessel and other vessels in the fleet. Unsafe periods are identified and their frequency of occurrences presented, but do not use weather conditions for channel design which exceed those which cause unsafe navigation at sea.

The following weather and hydraulic factors are needed for input to the navigation project design analysis.

1. Waves
2. Wind
3. Currents (Tidal and/or River)
4. Tides
5. Salinity
6. Visibility (Fog, Rain, Snow)
7. Ice

EM 1110-2-1615, Hydraulic Design of Small Boat Harbors (USACE 1984), refers to design conditions as follows:

Physical Data to Be Evaluated

The design of a small boat harbor project will require an analysis and evaluation of information on the following:

a. Weather
   1. Wind
   2. Waves
   3. Visibility (Rain, Smog, Fog, Snow)
   4. Ice

b. Weather
   1. Currents (Tidal, River, Seiche, Wave Generated)
   2. Sediment movement or long shore drift [*sic*]
   3. Type of bottom (Soft or Hard)
   4. Water depths and water level fluctuations
   5. Obstructions (Sunken Vessels, Abandoned Structures, etc.)
   6. Existing bridge crossings (Location, Type Clearance)

The factors listed above provide the basis for selecting the project design conditions. These design conditions must reflect weather and site conditions which are infrequently exceeded during the navigation season. Extreme weather conditions are to be evaluated and estimates of project damage presented.

A third example of design conditions is taken from EM 1110-2-2904, Design of Breakwaters and Jetties (USACE 1986).

The project design life and the degree of protection are required before design conditions can be selected. The economic design life of most breakwaters and jetties is 50 years. The degree of protection during the 50 year period should be selected by an optimization process of frequency of damages (both to the structure and the area it protected) when waves exceed the design wave.

Selection of a design vessel (ship or tow)—typically its beam, draft, and length, but also the propulsion system—dictates many design decisions.

Channels and basins must be designed to accommodate the maximum draft plus safe underkeel and side clearances. Channel curvature design must take into account the length of the design vessel and how maneuverable it will be under various current and wind conditions. Design vessel considerations are given for ASCE Manuals No. 50, Planning and Design Guidelines for Small Craft Harbors; No. 94, Inland Navigation: Locks, Dams and Channels; and No. 107, Ship Channel Design and Operation.

4.2 DESIGN ASSUMPTIONS

In concert with selection of design conditions, some design assumptions are needed before design can proceed. Design assumptions encompass the human factor, equipment failures, and emergency operation scenarios. Some examples of design assumptions are:

- Navigation Projects—Assume that vessels will be operated by sober, competent captains or pilots.
- Lock Design—Assume engine failure for a tow approaching a lock. The approach current should guide the unpowered tow safely into the lock or at least not sweep it into the spillway.
- Dam Design—Assume the powerhouse is shut down during a flood. Spillways should safely pass flood flows without contribution from the powerhouse.

4.3 RISK ASSESSMENT

An assessment of risk to life and property is needed for a project designed to function up to the initial set of design conditions and assumptions. If the risk (a combination of the probability of failure and consequences of failure) is not acceptable, then a project with a higher level of protection (a new set of design conditions and assumptions) needs to be evaluated for risk. This process is repeated until a project with an acceptable risk level is found. There are several types of risk to be considered in navigation projects. These risks are:

- Risk to public safety—Flooding from levee breaks, dam breaches, ferry accident, or leaks of dangerous cargos.
- Risk to users—Collisions and grounding of vessels.
- Risk to the economy—Interruptions of cargo movement due to waterway or lock shutdowns.
- Risk to the environment—Discharge of hazardous cargo or damage to habitat or migration, through pollution or loss of pool.

Risk is discussed further in Chapter 11.

4.4 ETHICAL CONSIDERATIONS

Some design conditions, such as selecting a 100-year event for design, are set as a matter of public policy or standard engineering practice. For example, a 50-year design life for structures has long been common practice. However, with many navigation locks continuing in use long after their design life [see "Report Card for America's Infrastructure–Navigable Waterways" (ASCE 2005)], it could be argued that engineers have a duty to include consideration of longer actual usage life spans under Canons 1 and 4 of the ASCE Code. This presents potential conflicts with standard practice, regulations, or even laws.

A more common ethical situation occurs in selection of the design vessel. Since new project design requires selection of a design vessel that will use the waterway in the future, substantial judgment must be exercised to determine which vessels of the future are likely to be accommodated. Overestimating the design vessel will produce a too-costly project, but underestimating it will reduce benefits—both affecting the efficiency and cost-effectiveness of the project. Such choices must be made consistent with the Canon 4 requirement to serve as faithful agents.

CHAPTER 5

CRITERIA FOR DESIGN OF PROJECT FEATURES

5.1 GENERAL

The dictionary defines a criterion as a "standard on which a judgment or decision may be based." Engineering criteria are normally the result of extensive performance history, model test results, and judgment of senior engineers' broad experience and perspective. The performance of project features such as locks, jetties, channels, etc. over a long period of time will expose weakness in design. Model tests will simulate structural performance during extreme conditions to show limits of survivability. These factors of prototype performance and model study results, when combined with engineering judgment, produce design criteria.

Design criteria, or guidance as it is called by the Corps of Engineers, is not usually intended to be inflexible. Criteria are intended to be a guide for design which can be modified if warranted. An example of the use of guidance criteria is presented in EM 1110-2-2602, Planning and Design of Navigation Locks (USACE 1995a) as follows:

> This engineer manual provides guidance and criteria for the planning, engineering layout, and design of navigation locks and appurtenant structures.
>
> This guidance will be followed in the design and layout of navigation locks, unless site-specific conditions or proposed innovative designs warrant deviations from the guidance contained in this manual. Deviations from this guidance should be approved by CECW-ED (Headquarters of the Corps of Engineers) and subsequently documented in design memoranda.

The latter statement is key to invoking flexibility in design criteria. If circumstances dictate that the design criteria should not be followed, a careful documentation of the reasons must be an integral part of the design notes so that checkers and reviewers can clearly understand both the rationale and the consequences of deviation.

Design criteria are usually imposed by the designing organization or the political jurisdiction of the project. Guidelines for choosing a set of criteria are:

- Know which criteria apply to your project and location.
- If none are mandated, use one of the national standards (Corps of Engineers, Coast Guard, etc.)
- Document the selection in your design notes.

5.2 CHANNELS

Some examples of design criteria that relate to safety, efficiency, or reliability for navigation channels are:

a. The channel is designed to allow safe and efficient transit of the design vessel under most weather conditions. ER 1110-2-1404, Hydraulic Deep-Draft Navigation Project Design (USACE 1981).
b. Channel depth. Channels in existing rivers must be consistent with connecting channel depths. Normal water depth in confined channels is to be a least 25% greater than the draft of the design vessel draft. ER 1110-2-1458, Hydraulic Design of Shallow Draft Navigation Projects (USACE 1998).
c. Channel alignment. Channels will usually follow the natural river course. ER 1110-2-1458, Hydraulic Design of Shallow Draft Navigation Projects (USACE 1998).
d. The channel design should permit passage of the design vessel, at the helm of a competent pilot or captain, under most weather conditions. ASCE Manual No. 107, Ship Channel Design and Operation (McCartney et al. 2005).

An example of a ship channel is shown in Fig. 5-1, and an inland barge channel in Fig. 5-2.

5.3 DAMS, INCLUDING SPILLWAYS AND POWERHOUSES

Stilling basins and downstream scour protections will be designed for the following conditions [ER 1110-2-1458, Hydraulic Design of Shallow Draft Navigation Projects (USACE 1998)]:

a. Uniform discharge through all spillway gates for a range of headwaters and tail waters expected during project life.

FIGURE 5-1. Detroit River Ship Channel. Courtesy of U.S. Army Corps of Engineers.

b. Single gate (each gate studied separately) full open with normal headwater and minimum tail water. This condition would assume incorrect gate operation. Minor damage to the downstream scour protection is acceptable as long as the integrity of the structure is not jeopardized. Single gate full open with above normal pool should also be considered. This would simulate several gates blocked by loose barges.
c. Single gate (each gate studied separately) open sufficiently wide to pass floating ice or drift at normal headwater and minimum tail water. No damage is acceptable for this condition.
d. Physical model studies are needed to verify final design powerhouse. Hydropower installations at new or existing navigation dams must be sited and operated so that adverse impact on navigation is minimized. Physical model studies are usually needed to select optimum powerhouse location and operation and determine necessary flow control structures.

An example of a high head navigation dam is shown in Fig. 5-3, and a low head navigation dam in Fig. 5-4.

FIGURE 5-2. Tennessee-Tombigbee Waterway, Divide Cut, Barge Channel. Courtesy of U.S. Army Corps of Engineers.

5.4 LOCKS

According to ER 1110-2-1458, Hydraulic Design of Shallow Draft Navigation Projects (USACE 1998),

> a. Lock and dam layout. The lock and dam layout will provide safe vessel transit through the range of river conditions expected during the navigation season. The design should allow a down bound tow to lose power in the upper approach and have the river currents float the tow safely behind the upstream lock guard wall.[1] A general physical model or ship simulator study will be used to optimize the lock and dam layout.

[1]A lock guard wall is a guide wall which lies between the lock approach and the spillway on the upstream side of the lock. A guide wall can be on either the spillway or land side of the approach and serves to create a safe straight-in approach to the lock chamber.

FIGURE 5-3. Ice Harbor Lock and Dam, Snake River, Washington. Lift 100 Feet. Courtesy of U.S. Army Corps of Engineers.

FIGURE 5-4. Melvin Price Lock and Dam, Mississippi River near Alton, Illinois. 24-Foot Lift. Courtesy of U.S. Army Corps of Engineers.

b. Lock design. Locks are to be designed for safe and rapid filling and emptying. Locks for barges are to have hawser loads (hawsers are the lines that hold the ship or tow in the lock chamber) at 4,536 kg (5 tons) or less. Hawser loads for ships can exceed this 4,536 kg (5 ton) limit; however, they must remain in a safe range for the lock mooring facilities. Physical model studies are needed to determine hawser loads. Standard lock sizes listed in EM 1110-2-1611 will be used. Lock floor elevations are set for safe and rapid filling with acceptable hawser loads. Lower lock sills will be as low as possible (0 to 0.9 m (0 to 3 ft) above floor) to allow safe and rapid tow entry and exit. The upper sill will have the same or greater clearance as the lower sill. (USACE 1998)

And according to EM 1110-2-2602, Planning and Design of Navigation Locks

a. Lock emergency closure. Emergency situations occur at navigation locks when a lock gate becomes inoperative in an open or partially open position while a head differential exists between the chamber and upper or lower pool. Although the cause may be mechanical failure, the more frequent cause is a navigation error that holds the gate partially open. Although no universally accepted definition of emergency closure exists, the required action is generally understood to be that a closure structure must be rapidly placed in flowing water under head differential. (USACE 1995a)

Examples of high head and low head locks are shown in Figs. 5-3 and 5-4.

5.5 SMALL BOAT HARBORS

Small boat harbor design criteria include requirements such as:

a. The acceptable wave height will depend on the vessel size and types of moorage (piers or anchorage). A 2 foot wave may be acceptable in moorage areas for large fishing vessels, where a 1 foot wave may be the maximum acceptable at a boat ramp. [EM 1110-2-1615 Hydraulic Design of Small Boat Harbors (USACE 1984)]
b. Moorage or Anchorage Area: Depth should accommodate draft, trim, wave action, low tide and a minimum of one-foot safety clearance. [EM 1110-2-1615, Hydraulic Design of Small Boat Harbors (USACE 1984)].
c. Small boat harbors should be located on the outside of river bends because the inside of bends will shoal. [See Fig. 5-5.]

An example of small boat harbor is shown in Fig. 5-6.

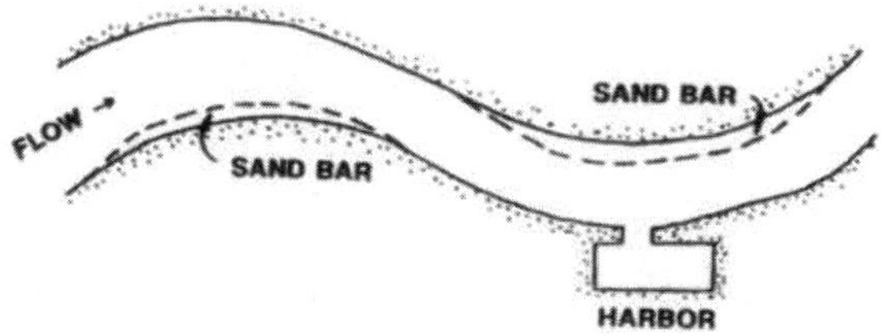

FIGURE 5-5. Site Harbors on the Outside of Channel Bends (McCartney et al. 2005).

5.6 JETTIES

The following criteria are taken from a Lessons Learned Appendix in ASCE Manual No. 107 (McCartney et al. 2005):

a. Parallel jetties are less prone to shoaling because of their configuration, which confines the ebb flow, raising ebb velocities and, thereby, flushing sediment seaward. [See Fig. 5-7.]

FIGURE 5-6. Small Boat Harbor, Cordova, Alaska. Courtesy of U.S. Army Corps of Engineers.

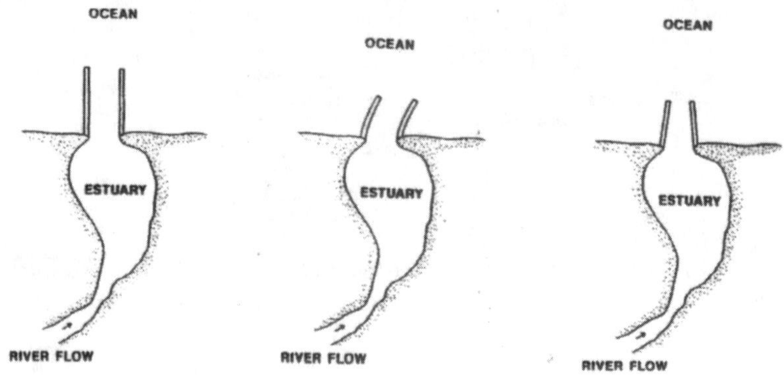

FIGURE 5-7. Straight, Parallel, Curved Parallel, and Arrowhead Jetties (McCartney et al. 2005).

b. Although curved jetties also can be designed to produce non-depositional velocities, flow concentrations on the outside of the curve can cause the undermining of the jetty and make a channel alignment difficult to navigate. [See Fig. 5-7.]
c. Entrance channels with arrowhead jetties frequently shoal rapidly because ebb flow is not confined enough to produce scouring velocities inside the jetties. [See Fig. 5-7.]
d. Jetties should be long enough to prevent littoral transport around the jetty ends and into the navigation channel. [See Fig. 5-8.]
e. Jetties should be sealed to prevent a significant portion of the littoral drift from passing through the jetty. [See Fig. 5-9.]

Examples of jetties are shown in Figs. 5-10 and 5-11.

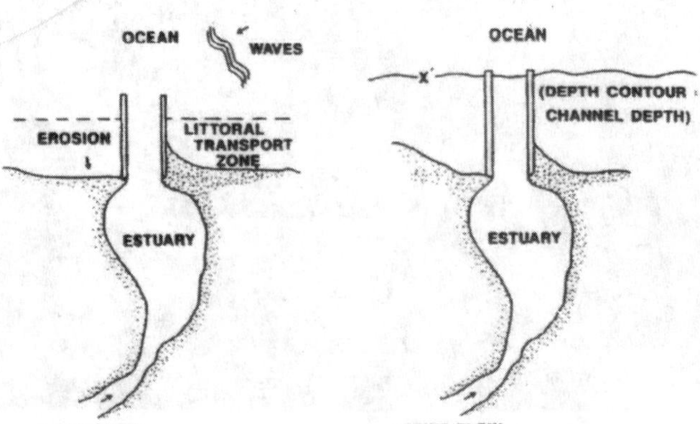

FIGURE 5-8. Jetty Length (McCartney et al. 2005).

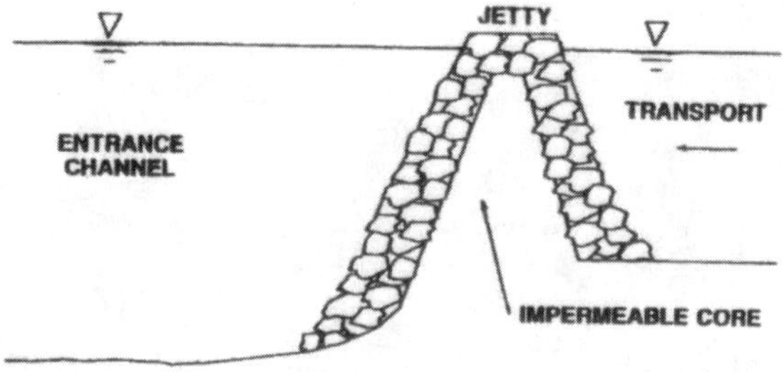

FIGURE 5-9. Jetty Sealing (McCartney et al. 2005).

5.7 ETHICAL CONSIDERATIONS

Ethical decision-making in design criteria arises when selecting the appropriate criterion for a particular design. For example, is choosing the minimum design criterion sufficient to meet the needs of both client and society? Does a criterion such as the hawser load limit listed in Section 5.4 add excessive costs or expose users to excessive risk in some situations? Legality does not automatically make a decision ethical, and neither does satisfying a specific design criterion if it is inappropriate to the situation

FIGURE 5-10. Jetties, Portage Lake Harbor, Onekama, Michigan. Courtesy of U.S. Army Corps of Engineers.

FIGURE 5-11. Jetties, Entrance to Umpqua River, Oregon. Courtesy of U.S. Army Corps of Engineers.

or if it is in conflict with another criterion or with the principles listed in Section 3.1.

Consider, for example, the jetty criteria of Section 5.6. Criteria (d) and (e) can limit littoral transport of sediment into the navigation channel, making the channel more reliable and reducing maintenance costs. However, in many sandy coast situations those criteria may produce structures that effectively block longshore sand transport and cause downcoast beach erosion, violating criteria for preserving beaches and protecting coastal property (principles of safety, social values, and sustainability).

CHAPTER 6
DESIGN PROCESS

6.1 DESIGN PROCESS

The design process is normally carried out in stages. It is initiated by a statement of project need. At this beginning stage, the design engineer or team takes a preliminary look to see if the project is feasible. This requires a determination of cost and benefits. If the project is deemed worthwhile, subsequent design phases refine and optimize the design which culminates in plans and specifications. The design phase is continued through construction to evaluate and accommodate any unforeseen conditions. An independent technical review of both computations and concepts is strongly recommended. This need for independent review has been recommended by the ASCE Hurricane Katrina Commission (ASCE 2006a). The design process combines hydraulic and weather conditions, vessel characteristics, design criteria, and social considerations to develop project features (channels, levees, dams, locks, etc.) that will provide a safe, efficient, reliable navigation project.

The steps in the design process are not necessarily the same for all waterways projects, but they generally include the following:

1. Review appropriate literature for current design practices and criteria.
2. Coordinate with users and other government agencies to determine their needs.
3. Conduct baseline surveys (pertinent physical and environmental data).
4. Select the design conditions and design criteria.
5. Develop concept design (usually with several alternative designs with annual costs and risk assessment).

6. Produce concept design with a recommended plan.
7. Refine recommended plan in final design stage.
8. Produce detailed plan which will include environmental sustainability evaluation and risk assessment.
9. Independent technical review at various stages of the design process.
10. Develop plans and specifications.
11. Evaluate design changes during construction phase.
12. Design deficiency evaluation if failure occurs during design life.

Some examples of variations on this generic design process follow.

6.2 DESIGN OF SMALL BOAT HARBORS

The Corps of Engineers EM 1110-2-1615, Hydraulic Design of Small Boat Harbors (USACE 1984) states:

> The design of small boat harbor projects requires an understanding of the problem, assembly and evaluation of all pertinent facts, and development of a rational plan. The design engineer is responsible for developing the design rationale and sufficient alternative plans so that the economically optimum plan is evident and the recommended plan is substantiated. Applicable Corps of Engineers guidance is considered in the design. Pertinent textbooks, research reports, or expertise from other agencies may be used as source information. The usual necessary steps leading to a sound plan are outlined below:
>
> 1. Review appropriate ER's, EM's, ETL's and other published information.
> 2. Assemble and analyze pertinent factors and environmental data.
> 3. Conduct baseline surveys.
> 4. Select a rational set of design conditions.
> 5. Develop several alternative layouts with annual costs.
> 6. Select an economically optimum plan.
> 7. Assess environmental and other impacts.
> 8. Develop a recommended plan.
> 9. Develop operation and maintenance plan.

6.3 DESIGN OF SHIP CHANNELS

ASCE Manual No. 107 (McCartney et al. 2005) gives a checklist for design of ship channels:

> The following checklist should be used during preliminary project design:
>
> 1. Review appropriate literature.
> 2. Consult with local port authority, pilot associations, and harbor terminal users.

3. Collect and analyze pertinent physical and environmental data.
4. Review appropriate local pilot or captain ship maneuvering strategy and evaluate existing navigation conditions.
5. Estimate volume and type of ship traffic and largest ships to be accommodated.
6. Estimate volume and type of commodity that will be moved.
7. Estimate amount, type and frequency of hazardous cargo (e.g., liquefied natural gas (LNG), ammunition, oil, radioactive material) movement, and evaluate special requirements.
8. Select and list the required project design operational conditions.
9. Select channel layout and alternative dimensions to be considered and determine advantages and disadvantages with annual costs.
10. Assess any adverse environmental and other impacts.
11. Define environmental mitigation needs and enhancement opportunities, especially beneficial uses for dredged material.
12. Review accident records for existing ship channels that are to be enlarged.
13. Security issues.

6.4 DESIGN OF APPROACH CHANNELS

The PIANC Supplement to Bulletin No. 95 (PIANC 1997) states:

Basic Definitions

Before considering the various stages in the design process, it is necessary to define some basic terms. Most important of these is the "approach channel."

An approach channel is defined as any stretch of waterway linking to berths of a port and the open sea. There are two main types:

- The seaway or outer channel, in open water.
- The main approach or inner channel which lies in relatively sheltered waters.

The channel normally terminates at its inner end in a swinging and/or berthing area which allows stopping and turning maneuvers to be made.

All sizes of approach channel are considered in this report; the problems of catering for small coasters in a small port may be as great as those for a large tanker at an oil terminal.

Stages of the Design Process

In this report approach channel design is considered to be a two-stage process consisting of:

- Concept Design
- Detailed Design

> As explained below, the methodology is based on the initial premise of a Design Ship, specified to represent the most testing ship expected to use the channel. In some cases, more than one Design Ship may be specified.
>
> In the Concept Design stage, initial estimates of the overall physical parameters of the proposed channel—width, depth and alignment—are determined from physical environment data and other information available at the outset. The Concept Design process is intended to be rapid in execution and not require excessive input data, so that alternative options (for trade-off studies) can be evaluated rapidly. The output physical parameters will be combined with proposals or assumptions on operational limits and aids to navigation.
>
> Detailed Design is a more elaborate process intended to validate, develop and refine the Concept Design, as regards both inputs and outputs. The methods used in Detailed Design commonly rely on computer models and therefore require more extensive and detailed input as well as needing proper judgment and experience in the interpretation of their output. The outputs of the Detailed Design may be subjected to further checking for acceptability by means of marine traffic analysis, risk analysis, and cost estimates. The results of these checks may lead to adjustments and a further cycle of detailed design.

The overall logic of the PIANC process is shown in Fig. 6-1.

6.5 DESIGN OF BREAKWATERS AND CLOSURE DAMS

d'Angremond and Roode (2001) provide this description of the design process for structures such as breakwaters:

> During the design process, one can also recognize certain phases that in some countries are related to the general conditions of contract between employer and consultant. Therefore the phases may vary from country to country. The contractual contents of each phase are subject to modification in the same way. A logical set of phases include:
>
> **Initiative**
>
> Formulation of the ultimate goals of the design object as part of the system.
>
> **Feasibility**
>
> Review of the system with respect to technical, economic, social, and environmental consequences and feasibility. Requirements are formulated on the component level.
>
> **Preliminary Design**
>
> Giving shape to the system on broad lines, including determination of the exact functionality of the components and definition of requirements at the element level.

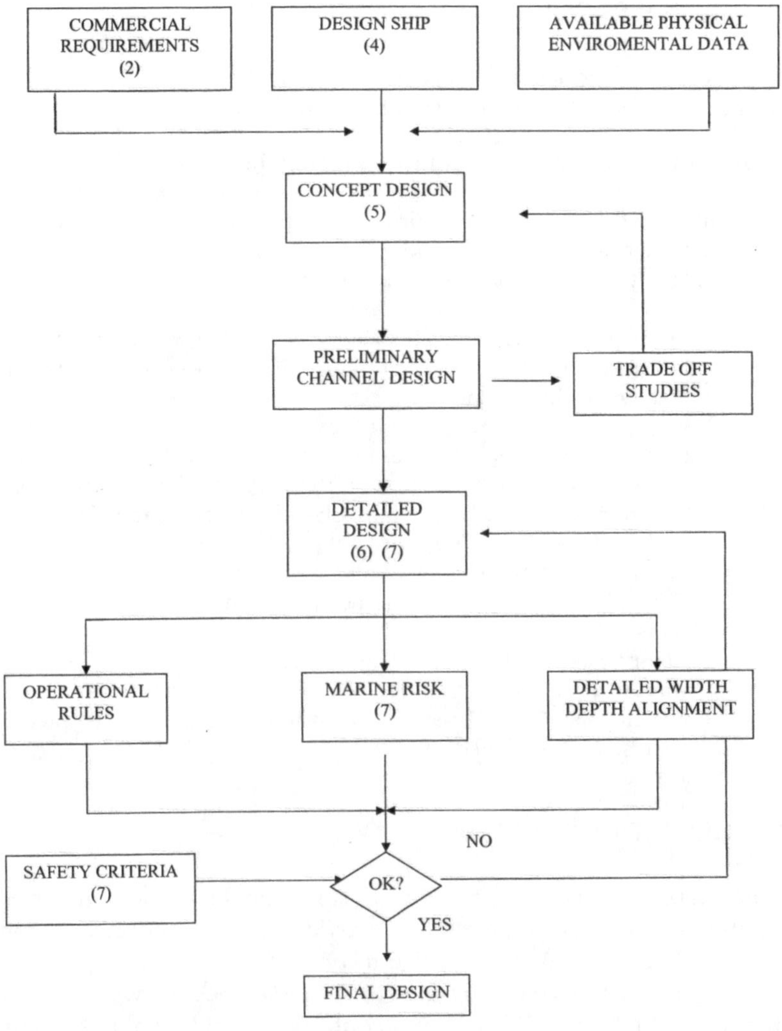

FIGURE 6-1. PIANC Approach Channel Design Procedure. Courtesy of International Navigation Association.

Final Design

Composition of a set of drawings and specifications for the system in which the final shape of the components is fixed and the functionality of the elements is determined.

Detailed Design

Composition of a set of drawings and specifications in which the final shape of the elements is fixed.

6.6 DESIGN OF LOCKS

The Corps of Engineers EM 1110-2-2602, Planning and Design of Navigation Locks (USACE 1995a) prescribes these steps:

> **Navigation lock planning principles and guidelines.** The objective of water resources planning is to contribute to national economic development (NED) consistent with protecting the environmental statutes, applicable executive orders, and other federal planning requirements. The planning process consists of the following steps:
>
> 1. Problem identification. This step specifies the water and related land resources problems and opportunities associated with the federal interest in navigation concerns.
> 2. Data gathering. Data gathering involves inventory, forecast, and analysis of water and land resource conditions within the planning area relevant to the navigation project problems and opportunities.
> 3. Alternative studies. This step involves formulation and evaluation of the effects of the alternative plans. The NED plan reasonably maximizes net NED benefits, consistent with the federal objective. Other alternative plans should be developed to address other concerns not listed in the NED plan.
> 4. Comparison of alternative plans. In this step, alternative plans and studies are compared in order to draw further conclusions.
> 5. Recommendations. Based on conclusions, a recommended plan is selected and presented.

6.7 INDEPENDENT REVIEW

An independent review provides a professional critique that is free of institutional bias or its appearance. A review team would include members with diverse areas of expertise to provide various viewpoints and they should be recognized experts in their fields. The reviews should start in the conceptual state of design and continue through final design. Independent boards of consultants have also been used during the construction phase when some complex or unique problem has surfaced.

ASCE Policy Statement 351, Project Peer Review (ASCE 2004) supports peer review and offers a peer review service for public agencies.

6.8 DESIGN DEFICIENCY EVALUATION

When a navigation project element fails to perform as intended or experiences frequent shutdowns or excessive maintenance, a review of the design is normally undertaken. This review can suggest structural or operational changes. However, for a "Design Deficiency" designation, the case must be made that criteria or standard practice was not followed at the time of the original design.

6.9 12 ACTIONS FOR CHANGE

The disastrous impact of the 2005 Hurricane Katrina on the Gulf Coast and the New Orleans area in particular was a wakeup call for the Corps, the nation, and the engineering profession. Exhaustive analysis by the Corps-commissioned Interagency Performance Evaluation Task Force and other teams of the performance of the Greater New Orleans Hurricane Protection System during Hurricanes Katrina and Rita yielded a number of lessons learned that pointed to the need for organizational changes to better serve the nation.

The "12 Actions for Change" (USACE 2006b) were developed from the analysis done by the Interagency Performance Evaluation Task Force, the American Society of Civil Engineers, the National Science Foundation-sponsored team, and Louisiana State University in the aftermath of Hurricane Katrina and earlier internal and external evaluations by the organization. These recommendations follow:

1. Employ an integrated comprehensive systems-based approach.
2. Employ risk-based concepts in planning, design, construction, and major maintenance.
3. Continuously reassess and update policy for program development, planning guidance, design, and construction standards.
4. Dynamic independent review.
5. Employ adaptive planning and engineering systems.
6. Focus on sustainability.
7. Review and inspect completed works.
8. Assess and modify organizational behavior.
9. Effectively communicate risk.
10. Establish public involvement risk reduction strategies.
11. Manage and enhance technical expertise and professionalism.
12. Invest in research and development.

These 12 initiatives are consistent with the design process examples given in this chapter but with renewed emphasis in certain areas. These general areas are:

- System-based approach—Look at the big picture to make sure that each element of the project is compatible with the desired project function.
- Identify risks and communicate risks to public and cost-sharing partners.
- Conduct independent reviews starting in the concept design phase.

- Inspect completed work—This is essential to determine whether what was constructed is in conformance with design. A final inspection should be preceded by ongoing inspections during the total construction phase. This returns the inspection mission to the responsible agency and eliminates contractor self-inspection and quality control.
- Enhance technical expertise—This entails development of in-house technical expertise and reduced use of design work outsourcing.
- Invest in research and development—Allow the organization to lead the way in innovative design.
- Focus on sustainability—Re-emphasize the initiative as contained in Canon 1 of the ASCE Code (see Chapter 7). An example of environmental concern is EM-1110-2-5026, Beneficial Uses of Dredged Material (USACE 1987).

These 12 actions are a reaffirmation of the traditional project development process and again focus on the goal of producing a "safe, efficient, and reliable" project. They require both an organizational commitment and a political commitment to fund projects and practices in conformance with ethical standards.

6.10 ETHICAL CONSIDERATIONS

The design process presents multiple opportunities for ethical dilemmas. As just one example, what should the design engineer do when she is concerned about project design decisions which she believes will compromise public safety? Article 3 of Canon 1 in ASCE's Code states:

> Engineers whose professional judgment is overruled under circumstances where the safety, health and welfare of the public are endangered, or the principles of sustainable development ignored, shall inform their clients or employers of the possible consequences.

Informing the client or employer must be done formally, both orally and in writing, so there is no opportunity for misunderstanding. If the client decides to proceed in spite of the engineer's concerns, the next step is a matter of discernment. If the engineer believes that laws are being broken, the whistle-blower rules (see Chapter 2) apply. For those cases where no laws are apparently broken but the action appears to be unethical, the discernment must include the engineer's decision as to whether or not they can, in good conscience, be a party to the action. The engineer must also consider that, as a Society member, it is their duty to report to ASCE any violation of the code that they have committed or observed, even if the act is legal. This situation is an example where the engineer should call the ASCE Hot Line for guidance. The call for guidance may be made anonymously but any further official action requires a signed written statement. Any action of ASCE (or other professional society) is internal, with the maximum penalty being expulsion of the member. However, ASCE may report the findings and action to the member's board of registration.

In a case known to the authors, an engineer was instructed to reduce the design schedule by 50%. He believed that the project could not be designed in half the time without compromising Canon 1 of the ASCE code, so he told his supervisor that the schedule could be physically met; however, it could not be done without adversely affecting the safety of the design and, if he were called to testify in court, he would be compelled to say that his advice had been rejected. The original schedule was kept.

CHAPTER 7
SUSTAINABLE DEVELOPMENT

7.1 GENERAL

> Sustainable water resource systems are those designed and managed to meet the needs of people living in the future as well as those of us living today. (ASCE 1998)

The earliest navigation projects were designed and constructed with minimal regard for depletion or destruction of natural resources. Design, construction, and operation of early ports and waterways (e.g., the ports at Alexandria, Egypt, about 330 B.C.E. and St. Augustine, Florida, about 1565 C.E., shown in Fig. 7-1) took advantage of water naturally deep enough to pass the ships of the day and employed at most small structures to provide safe anchorage. Whatever impact they had on their surroundings was ignored unless the port itself became threatened. Projects were often of too modest a scale for large environmental impacts, resources seemed abundant compared with their consumption, and science was inadequate to define all but the most catastrophic effects.

The earliest U.S. environmental law relating to navigation—the 1899 U.S. Rivers and Harbors Act—was designed to protect navigation, not the environment. The Act banned dumping of refuse or fill into waterways in order to protect shipping. Only in the 1960s was the prohibition on dumping of "refuse" interpreted as preventing pollution (USACE 2006c).

The dawn of modern environmental consciousness is often linked to two events: the 1962 publication of Rachel Carson's *Silent Spring*, and the first Earth Day in 1970. Carson's book forecast the loss of thousands of wild species and harm to humans if widespread, indiscriminate use of pesticides continued unabated. Despite furious criticism, accumulating evidence proved Carson right. In 1969 Sen. Gaylord Nelson called for a nationwide grassroots demonstration on behalf of the environment on

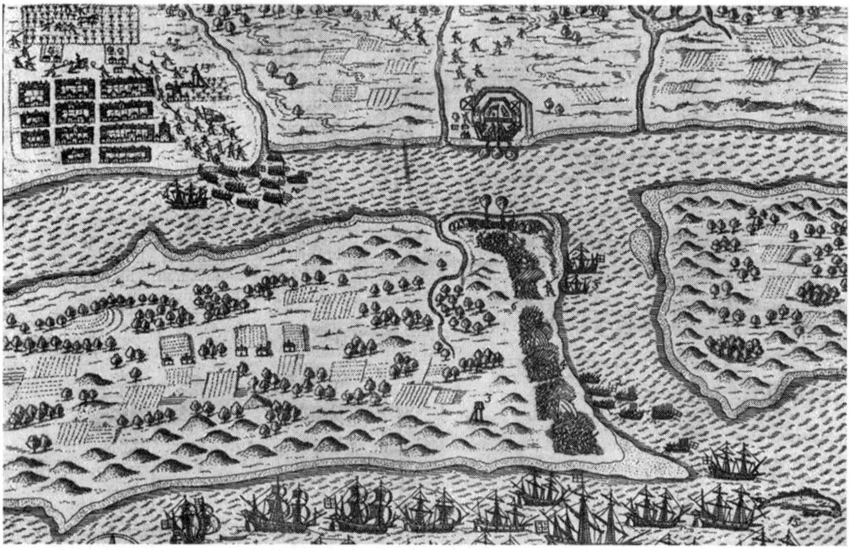

FIGURE 7-1. Port of St. Augustine, Florida. Detail from Expugnatio civitatis S. Augustini in America *by Theodor de Bry (1599). With kind permission of P. K. Yonge Library of Florida History, Department of Special & Area Studies Collections, George A. Smathers Library, University of Florida.*

April 22, 1970. He invited everyone to participate and more than 20 million people responded. Despite political opposition to the concept of environmental protection, and industrial and governmental skepticism about the threat of human activities to the environment, public opinion grew more and more supportive of environmental protection. Congress began passing laws designed to reduce environmental damage, including the Water Quality Act of 1965, the National Environmental Policy Act of 1969, and the Clean Water Act of 1972, which contributed to slowing the environmental degradation that Carson, Nelson, and others had warned of.

7.2 EVOLUTION OF DESIGN CRITERIA

Environmental laws and regulations gradually changed the practice of water resources engineering. At first, designers worked primarily to ensure that no laws or regulations were violated while designing for function and cost efficiency, as evidenced by design guidance wording such as ". . . while fully complying with all applicable environmental laws." Gradually the design emphasis changed to reflect avoidance of environmental harm, with guidance using phrases like ". . . minimize or eliminate adverse effects to the environment . . ." (USACE 1980). More recently the emphasis has evolved toward natural resources stewardship and the goal of sustainable development. For example, the Tennessee Valley Authority has adopted as one policy, "Practice responsible environmental stewardship of the Valley's natural resources," although it still retains the older mindset with another policy of, "Comply with environmental laws and regulations" (TVA 2006). Now, in addition to laws, regulations, and policy, the principles of sustainable development have been incorporated into codes of ethics, such as in Canon 1 of the ASCE Code (Chapter 3 and the Appendix). The Guidelines to Practice of the ASCE Code of Ethics includes the provision:

> Engineers should be committed to improving the environment by adherence to the principles of sustainable development so as to enhance the quality of life of the general public. (ASCE 2006b)

Despite this and similar statements in other engineering codes of ethics, sustainability remains as controversial as was basic environmental protection in the 1970s. A frequent criticism is that true sustainability is idealistic and impossible—any use of resources is bound to decrease the amount available to future generations. However, that criticism is no more valid than saying that we need not design and build for safety, since perfect safety is never achieved. Absolute sustainability can be an ideal goal that is balanced against other goals, such as economic development, or sustainability can become a design criterion if properly defined.

The ASCE definition given at the start of this chapter echoes ASCE Policy 418 (ASCE 2006c):

> Sustainable Development is the challenge of meeting human needs for natural resources, industrial products, energy, food, transportation, shelter and effective waste management while conserving and protecting environmental quality and the natural resource base essential for future development. The American Society of Civil Engineers (ASCE) recognizes the leadership role of engineers in sustainable development, and their responsibility to provide quality and innovation in addressing the challenges of sustainability. The ASCE Code of Ethics requires civil engineers to strive to comply with the principles of sustainable development in the performance of their professional duties. ASCE will work on a global scale to promote public recognition and understanding of the needs and opportunities for sustainable development.

7.3 DESIGN AND OPERATION CRITERIA FOR SUSTAINABILITY

Waterborne transportation is widely understood to be the most economical form of transport. Less well known is that it may have less impact on the air and water quality and natural habitat than equivalent highway, railway, pipeline, or air transport. For example, inland water transport of freight consumes much less oil (thus producing lower emissions) per ton-mile of transport than does highway transport and somewhat less than rail transport (Casavant 2000). Although reliable comparative metrics for other forms of resource consumption and degradation are untested, waterborne transport may also be more environmentally friendly in terms of habitat fragmentation, water pollution, and habitat destruction. Responsible decision-making on transportation policy and investment requires true intermodal performance metrics, including those for sustainability. An ASCE/UNESCO project has offered suggestions for measuring sustainability [Sustainability Criteria for Water Resource Systems (ASCE 1998)] that can be combined with intermodal transportation metrics (McAnally et al. 2004) to support decision-making in transportation.

Definitions and policies similar to these have been adopted by multiple organizations. Some, such as the International Navigation Association (PIANC), have translated sustainability policies into recommended design processes, and others, such as the U.S. Army Corps of Engineers, have expressed them in both design processes and criteria. For example, the Corps expresses a set of seven principals in its Environmental Operating Principles (USACE 2005a) and implements them through regulations and manuals.

In its Guidelines for Sustainable Inland Waterways and Navigation (PIANC 2003), PIANC recommends inland navigation project analyses that account for key physical and ecological processes, including:

- Morphologic processes
- Hydrologic processes
- Sedimentation balance
- Habitat provision
- Biological and chemical processes.

PIANC's recommended analysis procedure consists of seven steps (with numerous substeps):

1. Identify objectives and alternatives (established with all stakeholders) and agree on the reference situation.
2. Describe the waterways system and functions at the appropriate scale.
3. Are the waterway functions affected?
4. Are other uses affected or affecting the waterway?
5. Evaluate alternatives and select most preferable solutions with respect to effects on functions, other uses, and life-cycle costs.
6. Prepare the plan execution.
7. Monitoring and adaptive management.

In addition to the recommended analysis procedure, the cited PIANC report offers design alternatives that should be considered, such as ensuring adequate waterway depth by adapting vessel loads, training works, and flow regulation, as well as lock and dam structures and dredging.

USACE provides design processes and criteria through its Engineer Manual series (USACE 2006d), such as EM 1110-2-1202, Environmental Engineering for Deep-Draft Navigation Projects (USACE 2006e) and EM 1110-2-1607, Tidal Hydraulics (USACE 1991).

7.4 EXAMPLE OF ETHICAL CONSIDERATIONS

The USACE Houston-Galveston Navigation Channels Widening and Deepening Project provides one example of how sustainability (Canon 1 of the ASCE Code) can become an integral part of the design. The Houston-Galveston Navigation Channels Project, shown schematically in Fig. 7-2, will widen and deepen the Houston Ship Channel from 40 feet deep by 400 feet wide (12 meters deep by 122 meters wide) to 45 feet deep by 530 feet wide (14 meters deep by 160 meters wide) over a 53.5-mile (86-kilometer) distance from the Gulf of Mexico through Bolivar Roads, Galveston Bay, Buffalo Bayou, and a new container terminal at Barbours Cut (Fig. 7-3). During the planning and design phases of the project, an interagency working group led by the U.S. Army Corps of Engineers and including state and federal resource agencies and the Port of Houston deliberated on designs that would provide improved port access while

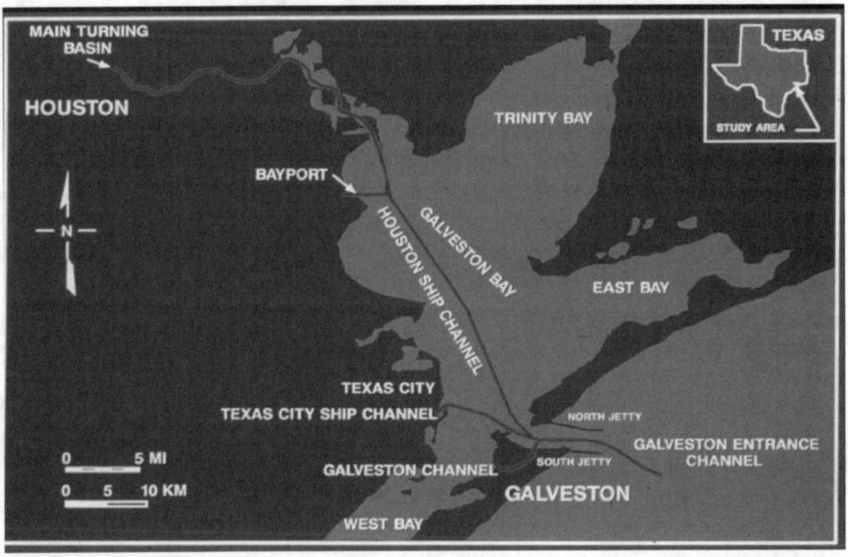

FIGURE 7-2. Houston-Galveston Ship Channels, Texas. Schematic. Courtesy of U.S. Army Corps of Engineers.

FIGURE 7-3. Barbours Cut Terminal in Houston, Texas. Courtesy of U.S. Army Corps of Engineers.

FIGURE 7.4 The Houston/Galveston Navigation Channel Project, Texas, Includes Construction of a Bird Island and More Than 4,000 Acres (1,620 Hectares) of Intertidal Marsh Habitat. Courtesy of U.S. Army Corps of Engineers.

preserving the valuable fisheries of the bay, providing habitat, and protecting water quality. In a process parallel to the seven-step PIANC recommendations, extensive observations, numerical modeling, and analyses of physical and biological processes were used to design the project so that functionality (e.g., navigability of the channel) was balanced with environmental enhancement. For example, over the project life of the channel more than 3,000 acres (1,200 hectares) of marsh, upland, and water bird habitat will be created under a beneficial uses of dredged material plan (Fig. 7-4 shows an example). Stakeholders described the process and final design to be very successful in terms of both project function and environmental quality (Jefts 2003).

This example demonstrates that a primary client (U.S. citizens) and individual clients and stakeholders (e.g., port, resource agencies, commercial fishers) may have multiple interests at stake in a project—economics, environmental quality, health and safety, etc.—that are sometimes in conflict with each other. Yet a dedicated, ethical approach to balancing interests and resolving differences can produce a project that each party can support.

CHAPTER 8

CORPS OF ENGINEERS MANAGEMENT OF WATERWAYS

8.1 OPERATIONS

Not very long ago there were various and often-segregated communities of practice, including the engineering community, the planning community, the operations community, numerous federal and state natural resource and regulatory agencies, along with non-government organizations (NGOs) and various users such as the recreational community and the commercial industry community. Each community had its own place, did its own job, and frequently made its own plans independently, often with minimal outside coordination or communication. There have been attempts to improve or enhance coordination efforts for more than 70 years; for example, the Fish and Wildlife Coordination Act of 1934, with amendments, provides the basic authority for the U.S. Fish and Wildlife Service's involvement in evaluating impacts to fish and wildlife from proposed water resource development projects (USFWS 2004).

Although few early attempts were successful, the advent of environmental laws of the late 1960s and the early 1970s, such as the National Environmental Policy Act (NEPA) and the Clean Water Act (CWA), were accompanied by an increased environmental awareness in the navigation communities, especially concerning impacts of channel maintenance operations to the waterway systems. Today this awareness, along with the increased need to improve efficiencies and reliability, as well as the development of new opportunities through legislation (various Water Resources Development Acts, for example), have resulted in numerous positive effects: Each community has begun to realize the need for and also the benefits of full, upfront, and continual interdisciplinary communication and coordination efforts for planning, design, construction, operations and maintenance (O&M), and monitoring of facilities and infrastructure along our waterways. In doing so they have discovered that the process of managing navigation projects has become more complex but also more consistent with Canon 1 of the ASCE Code: *Engineers shall hold paramount the safety, health, and welfare of the public . . . in the performance of their professional duties.*

8.2 PLANNING EFFORTS

Planning efforts are needed throughout the life of any project and not merely during the initial design, evaluation and construction phases. Once a project is constructed, planning is still required to ensure and verify compliance with project implementation plans, with natural resource and regulatory permits, laws, and regulations, as well as ongoing monitoring and reporting programs. In addition, projects may require major modifications that would necessitate additional coordination, planning, and design efforts. Projects may include the expansion of existing or development of new dredge material placement sites, or the development of staging areas for a major rehabilitation project at lock or harbor sites.

Planning results are usually better when they have been developed from a variety of perspectives, including the knowledge, skills, and insights of professionals from many of the natural, social, engineering, and environmental sciences along with other local experts. Using an interdisciplinary team is generally the best approach to the wide range of technical issues encountered in most studies; for example, refer to Chapter 2 of ER 1105-2-100, Planning Guidance Notebook (USACE 2000). This approach helps prevent violation of ASCE Canon 2: *Engineers shall perform services only in areas of their competence.*

The disciplines should be integrated so that each member of the team communicates their various viewpoints and works together to fashion plans that truly reflect a diversity of perspectives on the problems and opportunities that confront the planning area. An effective plan formulation process requires that the interdisciplinary team be involved in the planning process from the very beginning. Although the mix of disciplines required for a planning team varies from project to project, a team may include the following types of experts: archaeologists, attorneys, biologists, chemists, civil engineers, ecologists, economists, geographers, geologists, hydraulic engineers, hydrologists, landscape architects, planners, real estate specialists, and sociologists (USACE 2000). This list is not intended to exclude any discipline but, rather, to express the diversity that might be included. Moreover, local operations experts should be included in project development from the start of the planning through implementation, O&M, and monitoring.

All Corps planning studies are required to incorporate public involvement, collaboration, and coordination with their federal and non-federal partners and the public. This should be initiated during Step 1 of the planning process (Identifying Problems and Opportunities), as described in Section 2-3 of the Regulation (USACE 2000), and continue throughout the planning process. Involvement at the initial stage of the planning process not only helps to identify the problems and opportunities, but also extends an invitation to the public for continued involvement and a voice in the planning and decision-making process.

The nature of the planning study will determine who should be contacted. As a starting point, the following organizations, among others, should be considered: environmental/conservation groups; civic and neighborhood associations and community leaders; other federal, state, and local public agencies and entities; user groups; consumer and public interest groups; religious and ethnic groups; business groups, including small businesses and merchants; civil rights organizations; labor organizations; and organizations representing the handicapped, the elderly, low income segments of the population, the minorities, and the disadvantaged (USACE 2000).

One example of a planning study that has successfully incorporated a variety of effective coordination efforts is the Restructured Upper Mississippi River–Illinois Waterway System Navigation Study, also known as the Navigation and Environmental Sustainability Program (NESP). This study has included a great deal of public, stakeholder, interagency, and interdisciplinary involvement (USACE 2005b). This study includes newsletters, a program brochure, a communications network with stakeholders, public meetings, and mailings.

Uncertainty and variability are inherent in water resources planning. For example, there is uncertainty in projecting such factors as stream flows, population growth, and the demand for water. Therefore the consideration of risk and uncertainty is important in water resources planning. The planner's primary role in dealing with risk and uncertainty is to characterize, to the extent possible, the different degrees of risk and uncertainty and to describe them clearly so decisions can be based on the best available information. The planner should also suggest adjustments in design to reflect various attitudes of decision makers toward risk and uncertainty. If the planner can identify in qualitative terms the uncertainty inherent in important design, economic, and environmental variables, these judgments can be transformed into or assigned subjective probability distributions (USACE 2000).

Risk-based analysis, described further in Chapter 11, is defined as an approach to evaluation and decision-making that explicitly, and to the extent practical, analytically incorporates considerations of risk and uncertainty. Risk-based analysis shall be used to compare plans in terms of the likelihood and variability of their physical performance, economic success, and residual risks. A risk-based approach to water resources planning captures and quantifies the extent of risk and uncertainty in the various planning and design components of an investment project. The total effect of risk and uncertainty on the project's design and viability can be examined and conscious decisions made reflecting an explicit trade-off between risk and costs.

8.3 OPERATIONS AND MAINTENANCE

Once the primary planning, design, and construction of a project are completed, the O&M phase of a project begins. Studies and activities relative to the planning and construction phase may still be ongoing or initiated. For example, postconstruction monitoring may take place to determine whether or not project impacts are as predicted. Adaptive management actions may be required if actual conditions deviate significantly from those upon which the project was predicated. Additional activities such as development or finalization of mitigation management plans and acquisition of lands for compensation of unavoidable losses may also take place during the O&M phase (USFWS 2004). O&M practices are intended to retain the effectiveness of a project so that the project goals continue to be met and the design principles (Chapter 3) and underlying ethical requirements are honored.

O&M includes various maintenance scheduling, monitoring, and coordinating efforts that are carried out continually. Such activities may include the following:

- Development and revising of a chronological Standard Operating Procedure (SOP) for O&M
- Development of an emergency response plan
- Coordination efforts with users
- Coordination efforts with other (nearby or similar in nature) sites
- Weekly safety meetings and status reports
- Ongoing inspection during operations, preventive maintenance, and repair work
- Annual site inspections
- Maintaining, revising, and updating maintenance records
- Periodic inspections
- Staff assistance visits
- Development of a long-term maintenance plan
- Development of and revising the backlog of maintenance list or capital improvement list.

8.4 CHRONOLOGICAL STANDARD OPERATING PROCEDURE FOR O&M

An SOP for O&M includes a list of regular operations and maintenance items that are typically done for a project throughout any given year. It should be initiated as soon as the project is implemented. It should be considered a working document that could be revised or updated at any time. The SOP can be as short as one page or as detailed as the project manager deems necessary. The benefits of developing and using a chronological SOP include the following:

- Orienting new workers to daily project operations.
- Ensuring that no critical elements fail to be performed.
- Documenting to regulatory or resource agencies, or to planners, that critical work is scheduled and/or performed.
- Keeping all workers and staff on the "same page" of operations, required maintenance, and priorities.

An SOP may be developed for channel maintenance activities for upcoming dredging and dredged material placement activities, or for regulatory permit compliance activities, or for lock and dam (or other facility) operations and maintenance. Figure 8-1 shows a lock and dam site at Dresden Island Lock on the Illinois Waterway. In addition to operating the lock and dam system to pass traffic and regulate water levels, the onsite crews also perform routine maintenance of the lock and dam gates and valves and other components. They are also responsible for maintenance of the buildings, roads, and grounds, which includes landscaping

FIGURE 8-1. Dresden Island Lock and Dam Facility and Grounds. Courtesy of U.S. Army Corps of Engineers.

and security. This work is performed between regular lock and dam operations. In addition, the crews often need to continue regular O&M activities during extreme conditions such as high water, high flows, or ice, as well as during major maintenance and major rehabilitation projects. An SOP can help the crews focus consistently on regular O&M activities during extreme or chaotic situations.

8.5 DEVELOPMENT OF AN EMERGENCY RESPONSE PLAN

An emergency response plan is absolutely necessary to ensure continued reliable project performance, to prepare for rapid response to any unanticipated shutdown caused by mechanical breakdown or accident, and especially to provide formal communication lines among all affected parties in a timely manner. A coherent plan is essential for compliance with ASCE Code Canon 1.

There are three major types of emergency response plans. All three listed below involve a great deal of communication and coordination, whether on-site, within the corporate office, or among users and the public:

- USACE Emergency Management—The Corps continuously provides timely, effective, and efficient disaster preparedness, response, recovery, and mitigation projects and services on a nationwide basis to reduce loss of life and property damage under Department of Defense (DOD), USACE, Federal Emergency Management Agency (FEMA), and other agencies' authorities (USACE 2001). This includes a Corps Civil Emergency Management Plan (CEM) that focuses on (but is not limited to) Disaster Preparedness and Emergency Operations (response to floods, flood fights, and hurricanes, for example), Inspection and Rehabilitation Assistance, and Emergency Water Assistance. The CEM will not be addressed further in this section.
- Rapid Response for Emergency Repairs—The Corps plans and regularly meets within districts, adjacent districts, and division offices to ensure that emergency response plans are valid and implementable. Communication is vital in making these plans work. Each geographic region must have the capability to respond in a rapid timeframe to a variety of actions that have the potential to cause facility shutdowns. For example, heavy lifting equipment is required to change damaged miter gates at a lock site. Each Corps district has the capability for rapid response either using in-house equipment or using equipment from adjacent or nearby Corps districts, or using private contractor equipment as necessary. Because all types of this equipment are typically in heavy use, it is up to the facilities man-

agers to be apprised of equipment location and availability in the event a need arises. It is also advisable to have estimated times for arrival on site and knowledge of the availability of potential backup equipment. Figure 8-2 shows gate work being performed at Lock 19 on the Upper Mississippi River, Keokuk, Iowa.

- Waterways Action Plans (WAPs)—These plans are currently being developed for the western rivers by the U.S. Coast Guard in cooperation with the Corps, the river industry, and other participants. These plans will be used to guide the Coast Guard, the Army Corps of Engineers, and marine industry groups in taking appropriate and proactive measures to prepare for and react to extreme river conditions such as high water, high current, low water, ice, or other special circumstances (USCG 2006). It is the responsibility of the Coast Guard, the Corps of Engineers, and river industry representatives to meet and discuss changing conditions of a waterway system and to annually review and update the actions specified in the action plans. Each waterway is broken into different zones; each zone is delineated by river mile, and these zones are derived from a combination of reference gages, historical data, and known impact areas.

FIGURE 8-2. Lock 19 Major Maintenance Work. Courtesy of U.S. Army Corps of Engineers.

These plans will be consolidated into one concise plan, with annexes for each river system within the western rivers. There have been numerous action plans in the past but they have not included all of the western rivers; the plans had different terminology; and the plans did not fully address each of the possible river extremes. The WAPs will establish common framework for proactive and reactive efforts, ensure safety of life and navigation, provide protection of infrastructure, and prevent marine casualties during periods of high water, high velocity, low water, and ice.

Figure 8-3 gives an example of how high flows can affect navigation and infrastructure.

8.6 COORDINATION EFFORTS WITH USERS

Open communication and ongoing coordination between the project managers and operators, the users, and other stakeholders are vital to help keep each other aware of concerns regarding facility efficiencies (or lack thereof). Such communication would update affected parties regarding upcoming scheduled maintenance or rehabilitations that would require shutdowns, slowdowns, or unscheduled repairs that may add to delays. Working together helps to ensure that all concerns are heard and

FIGURE 8-3. Barge Caught in Dam by High Flows. Courtesy of U.S. Army Corps of Engineers.

priority issues are discussed along with potentials for delays to the navigation industry. One example of a successful coordinating committee is the Illinois River Carriers Association (IRCA). This committee is responsible for coordinating navigation activities along the Illinois Waterway. They meet with the Coast Guard and the Corps of Engineers, as well as other interested parties, five times per year. Because the Illinois Waterway is more than 300 miles (485 kilometers) in length, the IRCA rotates meeting locations between the Chicago area, Peoria, and St. Louis to cover the upper, middle, and lower portions of the waterway. This provides regular opportunities for users, stakeholders, and local agency representatives to participate in these coordination meetings without significant travel hardship. In addition, IRCA coordinates with other river committees and often will combine meetings in St. Louis with, for example, the River Industry Action Committee (RIAC). The RIAC is responsible for coordinating navigation activities along the upper and middle Mississippi River.

8.7 COORDINATION EFFORTS WITH OTHER SITES

This type of coordination provides opportunities for information exchange between facilities, between Corps districts, between Coast Guard offices, and other agency offices. Many of the waterway systems go beyond state, Corps district, or other agency regional boundaries. As a result, the need for regional coordination is very important to ensure continued efficient and reliable operations, as consistently as possible. Corps districts meet with adjacent districts on a regular basis to keep lines of communication open. For example, the St. Paul, Rock Island, and St. Louis districts meet annually to discuss channel maintenance issues and lessons learned. They also regularly meet with natural resource and regulatory agencies to update all parties on project(s) status, including funding and scheduling. Delays in progress are reported and potential alternative solutions are discussed.

Numerous other coordinating committees and organizations help to provide exchange of information. One example is the National Waterways Conference, Inc. (NWC). The purpose of NWC is to promote a better understanding of the public value of the American waterways system and to document the importance of farsighted navigation and water resources policies to a sound economy, industrial and agricultural productivity, regional development, environmental quality, energy conservation, international trade, defense preparedness, and the overall national interest.

8.8 WEEKLY SAFETY MEETINGS AND STATUS REPORTS

The Corps's safety manual (USACE 2003) requires weekly safety meetings for all field sites (also known as "tool box" safety meetings). These meetings give all workers an opportunity to hear any updates regarding a site, any potential hazards, and safety precautions. This also gives all workers the opportunity to express any concerns that need addressing. This may also be a good time to review previous work or accidents, identify what worked well and what did not work well, and discuss options to help ensure that what did not work well will not be repeated. This helps to keep safety and general work habits as higher priority and also helps to keep both workers and supervisors more accountable for their actions.

8.9 ONGOING INSPECTION DURING OPERATIONS, PREVENTIVE MAINTENANCE, AND REPAIR WORK

The above discussion concerns on-site lock operators and the work that is required of them in addition to locking boats, including regular maintenance and compliance with safety regulations. In addition, lock operators must continuously view components of the site during the regular operations to help maintain efficient operations and to help reduce delays to navigation. One example of this type of work involves the armor plating on the lock walls. Figure 8-4 shows significant damage to such plating. This type of damage may be caused by deteriorated concrete and/or by

FIGURE 8-4. Backlog of Maintenance Example: Damaged Wall Armor Plating. Courtesy of U.S. Army Corps of Engineers.

FIGURE 8-5. Backlog of Maintenance Example: Barge Damaged—Hit Wall Armor Plating. Courtesy of U.S. Army Corps of Engineers.

barge impact—newer barge repairs have frequently included an additional protection plate without chamfers, and these hard, square edges can compound a deteriorating condition. Figure 8-5 shows what can happen to a barge when it hits protruding metal. In this case the experienced lock operators were able to cut away most of the damaged plating with minimal delays to navigation. Traffic was allowed to continue unrestricted and the armor plating was scheduled for replacement. Ongoing inspection by on-site workers can help to identify minor problems and many times correct them before they develop into major repairs or major delays. Again, continual open communication between site personnel and the boat pilots can certainly help with exchange of information.

8.10 ANNUAL SITE INSPECTIONS

The project manager should meet *on-site* with the facility manager and the maintenance manager/foreman regarding needed maintenance work at each facility. This is relatively informal and should typically take no more than two to three hours, depending on the size and condition of the facility. On-site inspections of this type are vital to understanding site conditions and ensuring proper maintenance of a facility. The meeting may start with an After Action Review (AAR) of the previous season's required work list to discuss what work was not completed and why; what worked well during repairs and what did not; what work is still required (some items on the list may have been done with another repair crew or contractor, or by on-site personnel); and what lessons can be learned and used for upcoming work.

Once the review is completed, the group should discuss required work for the upcoming year, add items not completed the previous season, and revise the priority ranking, factoring (a) what items are most critical to ensure reliable site operations, and (b) what work can be accomplished using existing funding and/or equipment capability.

8.11 MAINTAINING, REVISING, AND UPDATING MAINTENANCE RECORDS

Record-keeping is one of the most important parts of site maintenance but has often been one of the more neglected, and it is difficult to perform regular maintenance on a site component or other piece of equipment without knowing when the previous maintenance was completed. Until recently, most site managers have used a card system to record any maintenance or repairs done on-site. There has typically been an individual card for each component (e.g., lock gate) or other piece of equipment on-site (e.g., backhoe). This system usually can be very effective but it is difficult to quickly review these types of maintenance records for any trends or patterns. For example, it could benefit the facility to know that an engine has been requiring additional maintenance or repairs during the past five years, or that a particular gate valve required new seals more often than is normal. It could also help to know whether the costs to maintain a specific component are changing. On-site personnel might be able to sense these changes and verify them by reviewing past records, but this is often not done because of time constraints or other more pressing items.

8.12 PERIODIC INSPECTIONS

The safety of dams and other infrastructure is a major concern of the Corps of Engineers, just as it has been since the Corps began building dams in the 1840s (USACE 2004). The purposes of the Corps's Dam Safety Program are to protect life, property, and the environment by ensuring that all dams and appurtenant structures are designed, constructed, and operated safely and effectively under all conditions. Accomplishing these purposes requires commitments to continually inspect, evaluate, and document the design, construction, operation, maintenance, rehabilitation, and emergency preparedness of each dam and the associated public information efforts.

Civil works structures whose failure or partial failure could result in loss of life or major damage to permanent structures, utilities, or transportation facilities shall be periodically inspected and evaluated to ensure structural stability, safety, and operational adequacy. Such inspection shall be at a frequency of a maximum of five years (after initial and second periodic inspection). An intermediate inspection of all or some of the features may be scheduled, if warranted. In addition, under the authority of ER 1130-2-530, Flood Control Operations and Maintenance Policies (USACE 1996a), the Corps shall participate in inspections of a sponsor-operated and -maintained structure (e.g., a local flood protection project) to ensure that the structure is conforming to the requirements of the Project Cooperation Agreement, the agreed-upon inspection program, and the operation and maintenance program.

8.13 STAFF INSPECTIONS AND STAFF ASSISTANCE VISITS

Staff inspections provide the Corps of Engineers military commander with specific, compliance-oriented feedback on functional areas or programs within the command (USDOA 2001). Staff inspections are compliance-oriented and focus on a single functional area or few related areas. Examples of staff inspections include safety, maintenance, physical security, and resource management inspections.

Staff Assistance Visits (SAVs) are not inspections; they are teaching and training opportunities designed to assist field sites in performing their mission. An SAV may include representatives from every district element; they should evaluate compliance with applicable policies and procedures, and the management of administrative, logistical, mobilization, operational, technical, and other applicable missions of each district element. Generally, the deputy district engineer will serve as team leader (e.g., Rock Island District). These site visits should be held at least once every 18 to 24 months. These inspections and SAVs provide another opportunity for improved communication, which results in greater potential for additional input that might not be addressed through other methods.

8.14 DEVELOPMENT OF A LONG-TERM MAINTENANCE PLAN

A long-term maintenance plan is vital to effectively operate and maintain a facility or structure throughout the life of the project. Long-term plans may cover any length but generally should be a minimum of five years. The purpose of a long-term plan is to list the requirements for reliable operation and maintenance, set priorities for the critical needs, and plan and schedule implementation through existing authority(s), funding, or equipment capabilities. These types of plans may include items that are often large enough—in either cost, equipment, or time needs—to exceed the regular O&M funding capabilities but too small to fit into the backlog of maintenance (backlog of maintenance items are typically more than $100,000). Figure 8-6 shows a leaking dam head gate that may fall into this category. These items tend to fall under deferred maintenance until subsequent and additional degradation require emergency repairs.

Information used for long-term planning may come from a large variety of functions already in place, including coordination efforts with users and nearby facilities, safety meetings and status reports, ongoing inspection and annual site inspections, keeping maintenance records updated, periodic inspections and SAVs, and updating and revising the backlog of maintenance lists (some deferred maintenance items may be incorporated into a backlog of maintenance items, for example). All of these different functions are closely interrelated.

A good example of a long-term navigation channel maintenance plan is the U.S. Army Corps of Engineers, St. Paul District, Channel Maintenance Management Plan (CMMP) (USACE 1996b). This CMMP is a comprehensive long-term plan for channel and harbor maintenance-related activities on various navigation projects in the St. Paul District. It identifies designated dredged material placement sites, describes a strategy for

FIGURE 8-6. Backlog of Maintenance Example: Leaking Dam Head Gate. Courtesy of U.S. Army Corps of Engineers.

placement site planning, discusses alternative channel maintenance techniques, and documents policies and procedures. Although long-term in nature, the plan is designed to accommodate new information and changes. The plan is periodically reevaluated based on factors such as changing regulations and authorities, economic or environmental conditions, and public opinion.

8.15 DEVELOPMENT AND REVISING THE BACKLOG OF MAINTENANCE OR CAPITAL IMPROVEMENT NEEDS

O&M and major rehabilitation programs are unable to adequately fund maintenance activities to ensure the navigation system operates at an acceptable level of performance. The backlog of maintenance contains maintenance, repair, and improvement needs for area infrastructure and projects. This type of list benefits the owner and the user by better identifying and documenting required work, ranking the work in order of critical priorities.

The U.S. Army Corps of Engineers, Mississippi Valley Division (MVD) created a Product Delivery Team (PDT) in 1996 to address this problem regionally for the entire Mississippi River and Illinois Waterway Locks and Dams systems. The regional approach should help improve inconsistencies and inefficiencies in addressing infrastructure concerns. The MVD Regional Backlog of Maintenance was valued at more than $1.29 billion in 2007 (USACE 2007).

Figures 8-7 and 8-8 show examples of backlog of maintenance items.

8.16 ETHICAL CONSIDERATIONS

Operating a waterway system requires that a multitude of tasks be performed with skill, diligence, and rigor, as described in this chapter on Corps of Engineers methods. Neglect of these management practices creates unsafe conditions and degrades system performance just as surely as do shoddy design practices, yet during times of budget constraints maintenance activities are often the first to be sacrificed. Canons 1 and 4 of the ASCE Code can thus be violated by acts of omission as well as commission. Although the funding problem is a political one, it has an engineering ethics component. ASCE has exercised its ethical responsibilities by means of a "report card" on America's infrastructure, calling attention to inadequately maintained waterborne transportation systems, including the navigable waterway system, which received a "D−" grade in 2005, down from a "D+" in 2001. The ASCE report (ASCE 2005) made three recommendations:

- Congress should amend the Inland Waterways Trust Fund Act of 1978 to allow all funds collected to be used for repair and construction

FIGURE 8-7. Backlog of Maintenance Example: Deteriorated Concrete on Dam Pier. Courtesy of U.S. Army Corps of Engineers.

FIGURE 8-8. Backlog of Maintenance Example: Deteriorating Dam Gate. Courtesy of U.S. Army Corps of Engineers.

of dams and locks. Congress should then appropriate the full fund balance each year to pay for the cost of rehabilitating the nation's oldest locks. The government needs to set a priority system for restoring locks that have outlasted their design lives, with an initial focus on all locks built in the 19th century. The current federal budget process does not differentiate between expenditures for current consumption and long-term investment. This causes major inefficiencies in the planning, design and construction process for long-term investments.

- In the interim, Congress must appropriate the full amount in the Inland Waterway Trust Fund to begin reducing the maintenance backlog.
- The American Society of Civil Engineers (ASCE) supports the creation of a federal capital budget to create a funding mechanism that would help reduce the constant conflict between short-term and long-term maintenance needs. This would help to increase public awareness of the problems and needs facing this country's physical infrastructure, and would help Congress to focus on specific programs devoted to long-term growth and productivity.

For an individual engineer working on a single project, the ethical dilemmas may be more subtle. Certainly the individual should implement the kind of methodical operation procedures described here and immediately call to management attention any deficiencies in the project that compromise safety or function. However, one year of deferred maintenance may not by itself significantly compromise project integrity. In this and similar situations, discernment must be carefully exercised to avoid being either overly alarmist or apathetic. To fully meet Canon 1, the profession must keep these issues visible to the public and proper officials.

CHAPTER 9

U.S. COAST GUARD CONTRIBUTIONS TO WATERWAYS[1]

9.1 GENERAL

The U.S. Coast Guard carries out numerous safety missions and tasks, including port safety and security, waterways management, and commercial vessel safety. It is responsible for a safe, efficient, and navigable waterway system to support domestic commerce, international trade, and military sealift requirements for national defense. The services that the Coast Guard provides include long- and short-range aids to navigation; charting; tide/current/pilotage information through "Notices to Mariners"; vessel traffic services; domestic and international icebreaking and patrol services; technical assistance and advice; vessel safety standards and inspection; and bridge administration standards and inspection. These services can be consolidated into five fundamental roles:

- Maritime Mobility
- Maritime Safety
- Maritime Security
- National Defense
- Protection of Natural Resources.

Many Coast Guard missions benefit more than one of its roles. For example, whereas the Aids to Navigation mission primarily supports the service's Maritime Mobility role by facilitating the movement of people and goods, the system of aids also supports the Coast Guard's role of maintaining maritime safety and protecting natural resources by preventing accidents.

---

[1]This chapter consists of extracts from ASCE Manual No. 107, Ship Channel Design and Operation (McCartney et al. 2005) and USCG Publication 1 (USCG 2002).

9.2 MARITIME MOBILITY

The U.S. Marine Transportation System facilitates America's global reach into foreign markets and engagement in world affairs, including protection of U.S. national interests through a national and international regulatory framework governing trade and commerce. This system includes the waterways and ports through which most of America's foreign and domestic freight moves each year, as well as the intermodal links that support economic and military security. It also includes international and domestic passenger services, commercial and recreational fisheries, and recreational boating. The Coast Guard's primary missions for providing a safe and efficient marine transportation system include:

- Aids to Navigation
- Notice to Mariners
- Icebreaking
- Bridge Administration
- Waterways Management/Vessel Traffic Service.

9.2.1 Aids to Navigation

The waters of the United States and its territories are marked to provide safe navigation by the U.S. Aids to Navigation System. This system employs a simple arrangement of colors, shapes, numbers, and light characteristics to mark navigable channels, waterways, and obstructions.

Aids to Navigation provide a vessel operator with the same type of information drivers get from street signs, stop signals, road barriers, detours, and traffic lights. These aids include lighted structures, beacons, day markers, range lights, fog signals, landmarks, as well as floating buoys. Each aid has a purpose and helps in determining location, how to get from one place to another, or how to stay out of danger. Figure 9-1 shows the placement of a channel marker buoy.

The U.S. Aids to Navigations System is intended for use with nautical charts, one of the most important tools used by vessel operators for planning trips and safely navigating waterways. Such charts show the nature and shape of the coast, buoys and beacons, depths of water, land features, directional information, marine hazards, and other pertinent information.

9.2.2 Notice to Mariners

The Coast Guard has statutory and treaty obligations to make navigation information available to the public. Local Notices to Mariners (LNMs) are the primary means for communicating information pertaining to individual Coast Guard districts. LNMs, which are available free of charge,

FIGURE 9-1. U.S. Coast Guard Cutter Acacia *Placing Channel Marker Buoy near Chicago. Courtesy of U.S. Coast Guard.*

provide important safety information that is not available anywhere else. LNMs appear on the Coast Guard Navigation Center's Web site (http://www.navcen.uscg.gov/lnm/default.htm) and include such information as submerged obstructions, missing or malfunctioning buoys, etc.

9.2.3 Icebreaking

For decades the Coast Guard has provided both domestic and international icebreaking services. Section 2 of Title 14 of the U.S. Code requires the Coast Guard to operate icebreaking facilities on domestic and international waters. In 1965 the Coast Guard and the Department of the Navy signed a Memorandum of Agreement that requires the Coast Guard to maintain and operate all U.S. icebreakers in wartime as well as undertake seasonal deployments to the Arctic and Antarctic in support of national interests.

Domestic icebreaking operations are performed on U.S. navigable waters in support of national and international maritime transportations, commerce, and safety. Geographically, domestic icebreaking is conducted in two regions: on the East Coast from Maine to Virginia, and on the Great Lakes. The Coast Guard's fleet of ice-capable ships includes both icebreaking cutters and buoy tenders. In domestic waterways, the Coast Guard conducts icebreaking operations to keep certain shipping routes and ports open during the winter to meet the demands of commerce. The Coast Guard responds to a vessel operator's requests for assistance if they are disabled or stranded in ice-covered waters. Figure 9-2 shows an icebreaker.

9.2.4 Bridge Administration

In 1967 the Bridge Program was transferred from the Corps of Engineers to the Coast Guard. The Coast Guard is responsible for approval of the location and plans of bridges and causeways constructed across navigable waters of the United States. In addition, the Coast Guard is respon-

FIGURE 9-2. Icebreaker, U.S. Coast Guard Cutter Healy. *Courtesy of U.S. Coast Guard.*

sible for approval of the location and plans of international bridges and the alterations of bridges found to be unreasonable obstructions to navigation. Any bridge that connects the United States with a foreign country is referred to as an "international bridge."

9.2.5 Waterways Management/Vessel Traffic Service

The Coast Guard has a statutory responsibility under the Ports and Waterways Safety Act of 1972 (PWSA), Title 33 USC 1221, to ensure the safety and environmental protection of U.S. ports and waterways. The PWSA authorizes the Coast Guard to ". . . establish, operate and maintain vessel traffic services in ports and waterways subject to congestion." It also authorizes the Coast Guard to require the carriage of electronic devices necessary for participation in the Vessel Traffic Service (VTS) system. The purpose of the act was to establish good order and predictability on U.S. waterways by implementing fundamental waterways management practices.

The VTS system at each port has a Vessel Traffic Center that receives vessel movement data from the Automatic Identification System (AIS), surveillance sensors, other sources, or directly from vessels. Meteorological and hydrographic data are also received at the Vessel Traffic Center and disseminated as needed. AIS technology relies on global navigational positioning systems (GPSs), navigation sensors, and digital communication equipment operating according to standardized protocols (AIS transponders) that permit the voiceless exchange of navigation information. AIS transponders can broadcast vessel information such as name or call sign, dimensions, type, GPS position, course, speed, and navigation status. This information is continually updated and received by all AIS-equipped vessels in the vicinity. An AIS-based VTS reduces the need for voice interactions, enhances the ability to navigate, improves situational awareness, and assists in the performance of duties, thus reducing the risk of collisions. Figure 9-3 shows the location of current Coast Guard VTS areas.

9.3 MARITIME SAFETY

One of the basic responsibilities of the U.S. government is to protect the lives and safety of Americans. In the maritime realm, the lead responsibility falls to the Coast Guard, which is part of the Department of Homeland Security. In partnership with other federal agencies, state and local governments, marine industries, and individual mariners, the Coast Guard preserves safety at sea through a focused program of prevention, response, and investigation.

9.3.1 Prevention

Safety prevention activities include developing commercial and recreational vessel standards, enforcing compliance with these standards,

FIGURE 9-3. Current U.S. Coast Guard VTS Areas. Courtesy of U.S. Coast Guard.

licensing commercial mariners, operating the International Ice Patrol to protect ships transiting the North Atlantic shipping lanes, and educating the public. The Coast Guard develops operating and construction criteria for many types of vessels, from commercial ships to recreational boats. The Coast Guard is America's voice in the International Maritime Organization (IMO), which promulgates measures pertaining to improving shipping safety, pollution prevention, mariner training, and certification standards. The Coast Guard is the agency primarily responsible for developing domestic shipping and navigation regulations.

Navigation and shipping regulations are published in Chapter I of Titles 33 and 46, Code of Federal Regulation (CFR). These regulations provide detailed guidance for the design and operation of inspected vessels, and establish minimal requirements for un-inspected vessels.

The Coast Guard ensures compliance with safety regulations in many ways. Members of the Coast Guard inspect U.S. flag vessels and mobile offshore drilling units and marine facilities; examine foreign-flag vessels based on the potential safety and pollution risk they pose; review and approve plans for vessel construction, repair, and alteration; and document and admeasure U.S. flag vessels. The Port State Control program is aimed at eliminating substandard foreign-flagged vessels from U.S. ports and waterways. Port State Control is a key element in the safety enforcement program because 95% of large passenger ships and 75% of cargo ships operating in U.S. waters are foreign-flagged.

9.3.2 Response (Search and Rescue)

Mishaps will occur despite the best prevention efforts. As the lead agency for maritime search and rescue (SAR) in U.S. waters, the Coast Guard coordinates the SAR efforts of sea and airborne Coast Guard units as well as those of other federal, state, and local responders. In addition, they also leverage the world's merchant fleet to rescue mariners in distress around the globe through the Automated Mutual-Assistance Vessel Rescue (AMVER) system.

The statutory authority for the Coast Guard to conduct SAR missions is contained in Title 14, Sections 2, 88, and 141 of the U.S. Code. The code states that the Coast Guard shall develop, establish, maintain, and operate SAR facilities; may render aid to distressed persons; and protect and save property on and under the high seas and waters subject to the jurisdiction of the United States. These waters generally include all navigable water subject to the jurisdiction of the United States but also include international waters stretching far into the Atlantic and Pacific oceans and the Gulf of Mexico.

The mission and purpose of the Coast Guard's SAR Program is to prevent death or injury to persons and loss or damage to property in the marine environment. SAR functions and the hierarchy of response can be broken into two parts:

- Search—An operation normally coordinated by a Rescue Coordination Center (RCC), Rescue Sub Center (RSC), or sector command, using available and appropriate personnel, facilities, and resources to locate persons or property in distress.
- Rescue—An operation with the primary purpose of retrieving persons in distress and delivering them to a place of safety. This may include providing for certain medical care or other critical needs. Rescue operations also may be performed for the purpose of preventing or mitigating property loss or damage. However, missions shall not normally be performed for the purpose of salvage or recovery of property when those actions are not essential to the saving of life. Beneficial secondary consequences of a rescue operation may be to prevent environmental damage or remove hazards to navigation, but these are not considered part of the rescue operation's objective.

The rescue of persons in distress is the highest-priority SAR mission. Missions *solely* for saving property or for other purposes such as preventing environmental damage will always give way to saving a person's life.

9.3.3 Casualty Investigations

An important purpose of marine casualty investigations is to obtain information to prevent similar casualties, as far as is practicable. It is necessary to determine the causes of casualties as precisely as possible so that factual information will be available for program review and statistical studies. It is not sufficient to know only how a casualty occurred; it also must be clear why it happened. Based on this information, the Coast Guard may develop appropriate corrective measures, regulations, and standards of safety. In addition, legislation for marine safety may be recommended, if needed. An equally important purpose of these investigations is the determination of whether there is any evidence of violation of law or regulation; any basis for the institution of civil penalty action under any of the laws administered by the Coast Guard; or suspension and revocation (S&R) proceedings under 46 U.S.C. 7703.

The Coast Guard has the jurisdiction to investigate the following:

- A marine casualty or other accident involving any vessel on the navigable waters of the United States, or involving U.S. vessels wherever they may be.
- An incident involving the destruction of, or damage to, any bridge or other structure on or in the navigable waters of the United States, or any land structure or shore area immediately adjacent to those waters.
- An incident involving a major fire, an oil spill, or any injury occurring as a result of operations conducted pursuant to the Outer Continental Shelf Lands Act (OCSLA), including allegations of unsafe working conditions or violations of safety regulations.
- Water pollution by oil or other hazardous substance or the threat thereof to the "waters of the United States" (anywhere in the hydrologic chain).
- Acts of misconduct, incompetence, negligence, unskillfulness, or willful violation of law committed by any licensed, certified, or documented individual.
- Boating accidents.
- Casualties or accidents that occur to any component of a deep-water port.

The primary purpose of an investigation is to ascertain the cause(s) of an accident, casualty, or personnel misbehavior to determine whether remedial measures should be taken, and to determine whether any violation of federal law or regulation has occurred. It should be clearly understood that the Coast Guard does not conduct investigations to determine civil liability in disputes between private litigants. Rather, its investigations are a means to promote safety of life and property and to protect the marine environment. Typical of a casualty investigation is the 2003 Dutch ship *Stellamare* incident where three crew members were killed. This accident is shown in Fig. 9-4.

FIGURE 9-4. Ship Stellamare *Sunk at Albany, New York. Courtesy of U.S. Coast Guard.*

9.4 MARITIME SECURITY

Maritime law enforcement and border control are the oldest of the Coast Guard's numerous responsibilities, dating back to its establishment as the Revenue Marine in 1790. Congress established the Revenue Marine specifically to patrol the coasts and seaports to frustrate smuggling and enforce the customs laws of the fledgling republic. The Coast Guard's maritime law enforcement role and the task of interdicting ships at sea provide the foundation on which its much broader and complex present-day mission set has been built. Maritime security missions include:

- General Maritime Law Enforcement
- Drug Interdiction
- Alien Migrant Interdiction
- Exclusive Economic Zone (EEZ) and Living Marine Resource Law/ Treaty Enforcement.

9.4.1 General Maritime Law Enforcement

As the nation's primary maritime law enforcement service, the Coast Guard enforces or assists in enforcing federal laws, treaties, and other international agreements on the high seas and waters under U.S. jurisdiction. They possess the authority to board any vessel subject to U.S. jurisdiction to make inspections, searches, inquiries, and arrests. The Coast Guard wields extraordinarily broad police power primarily to suppress violations of drug, immigration, fisheries, and environmental laws. No other U.S. armed service or federal agency possesses this combination of law enforcement capabilities and responsibilities together with the legal authority to carry them out.

The Coast Guard's ability to fulfill its roles (i.e., saving lives and property at sea; protecting America's maritime borders and suppressing violations of the law; protecting the marine environment; providing a safe, efficient marine transportation system; and defending the nation) makes the Coast Guard truly a unique instrument of national security.

9.4.2 Drug Interdiction

As the designated lead agency for maritime drug interdiction under the National Drug Control Strategy and the co-lead agency with the U.S. Customs Service for air interdiction operations, the Coast Guard defends America's seaward frontier against a virtual torrent of illegal drugs. For more than two decades Coast Guard cutters and aircraft, deployed off South America and in the transit zone, have intercepted cocaine, marijuana, and other illegal drugs that otherwise would have found their way to American streets. Figure 9-5 shows a drug interdiction patrol off Florida.

9.4.3 Alien Migrant Interdiction

Coast Guard alien migrant interdiction operations are also law enforcement missions with a significant humanitarian dimension. Migrants typi-

FIGURE 9-5. U.S. Coast Guard Cutter Gallatin *and Helicopter on Drug Interdiction Patrol Off Florida. Courtesy of U.S. Coast Guard.*

cally take great risks and endure significant hardships in their attempts to flee their countries and enter the United States. In many cases, migrant vessels interdicted at sea are overloaded and unseaworthy, lack basic safety equipment, and are operated by inexperienced mariners. The majority of alien migrant interdiction cases handled by the Coast Guard actually begin as SAR cases, once again illustrating the interwoven nature of the Coast Guard's roles and missions. Between 1980 and 2000 the Coast Guard intercepted 290,000 migrants, mostly from Cuba, the Dominican Republic, the People's Republic of China, and Haiti.

9.4.4 Exclusive Economic Zone and Living Marine Resource Law/Treaty Enforcement

In 1976 Congress passed what is now known as the Magnuson-Stevens Fishery Conservation and Management Act. By creating an Exclusive Economic Zone (EEZ), this act pushed out the U.S. maritime border to 200 nautical miles. In the years that followed, international fisheries agreements went even farther, extending U.S. jurisdiction to high seas areas beyond the EEZ. Today the Coast Guard patrols these areas as well as the EEZ—where they focus primarily on maritime boundary areas such as the U.S./Russian Convention Line in the Bering Sea—to uphold U.S. sovereignty and protect America's resources.

9.5 NATIONAL DEFENSE

Throughout American history, the Coast Guard has served alongside the U.S. Navy in critical national defense missions, beginning with the Quasi War with France in 1798, through the Civil War, World Wars I and II, to the Vietnam War and the Persian Gulf War. A 1995 agreement between the Secretaries of Defense and Transportation assigned the Coast Guard five specific national defense missions in support of the Unified Commanders in Chief (CINCs) in addition to their general defense operations and polar icebreaking duties. These missions (i.e., maritime interception operations; military environmental response operations; port operations, security, and defense; peacetime military engagement; and coastal sea control operations) require the Coast Guard to execute essential military functions and tasks in support of joint and combined forces in peacetime, crisis, and war.

9.6 PROTECTION OF NATURAL RESOURCES

The Coast Guard's protection of natural resources role dates to the 1820s when Congress required the Revenue Marine to protect federal stocks of Florida live oak. As the exploitation of the nation's valuable marine resources—whales, fur-bearing animal, and fish—increased, the Coast Guard was given the duty to protect these resources as well. Today, with the U.S. EEZ supporting commercial and recreational fisheries worth more than $30 billion annually, the Coast Guard serves as the primary agency for at-sea fisheries enforcement. This role has expanded over the last few decades to include enforcing laws intended to protect the environment as a public good. As a result, the Coast Guard now actively protects sensitive marine habitats, marine mammals and endangered marine species, and enforces laws protecting U.S. waters from the discharge of oil and other hazardous substances.

The Coast Guard conducts a wide range of activities (e.g., education and prevention, enforcement, response and containment, and recovery) in support of its primary environmental protection mission areas: maritime pollution enforcement, offshore lighting zone enforcement, domestic fisheries enforcement, and foreign vessel inspection. They are usually the first responders to environmental disasters on the seas and are typically the lead agency for any ensuing response effort. Under the National Contingency Plan, Coast Guard Captains of the Port (COTPs) are the predestinated Federal On Scene Coordinators (FOSCs) for oil and hazardous substance incidents in all coastal and some inland areas. The FOSC is, in reality, the president's designated on-scene representative. As such, the FOSC is responsible for forging a well-coordinated and effective response operation involving a diverse set of government and commercial entities in many emotionally charged and potentially dangerous emergency situations.

9.6.1 Pollution Response

The Coast Guard's concerns extend to pollution and threats of pollution in the coastal zone. This zone includes U.S. waters subject to the tide, U.S. waters of the Great Lakes, specified ports and harbors on inland rivers, and the contiguous zone and waters on the high seas out to 200 nautical miles. There are four elements involved in assessing discharges and releases to ensure appropriate response:

- Preventing spills whenever possible.
- Ensuring that responsible parties clean up discharges of oil and releases of hazardous substances.
- Mitigating the effects of spills that do occur.
- Reducing the potential for spills or operational discharges outside U.S. waters from entering U.S. waters or fouling U.S. coastlines.

These elements are considered in all cases of pollution or threatened pollution that arise from deep-water ports or outer continental shelf activities.

To respond to major discharges, the Coast Guard commandant has established the National Strike Force (NSF). This consists of teams that are pre-positioned on the Atlantic, Gulf, and Pacific coasts to assist on-scene coordinators (OSCs) of federal response activities. Strike teams are at the following locations: GULF Strike Team, Mobile, Alabama; NATIONAL Strike Force, Elizabeth City, North Carolina; PACIFIC Strike Team, Novato, California; and ATLANTIC Strike Team, Fort Dix, New Jersey. In addition, the NSF has assisted foreign governments on request in major international pollution cases.

The Coast Guard operates the National Response Center (NRC) around the clock to receive notification of pollution incidents and ensure that information is passed to the predesignated Coast Guard or Environmental Protection Agency (EPA) OSC for response. The NRC provides a toll free number (800-424-8802) for making pollution reports from anywhere in the United States.

9.6.2 Enforcement

The federal responsibility for the removal of discharged oil and hazardous substances is shared by the Coast Guard and the EPA. The federal agencies cannot require anyone to clean up an oil spill; however, if federal funds are used for the cleanup, the costs (up to certain limits) are passed to the owner or operator of the discharging vessel or facility or, under certain circumstances, to the person causing the discharge.

9.7 ETHICAL CONSIDERATIONS

Ethical situations in Coast Guard activities are substantially similar to those of any engineer working in navigation project operations. Many issues will center on adequate funding and staffing to conduct the agency's missions, as, for example, in maintaining Aids to Navigation. Members of the Coast Guard are governed by an internal set of ethical guidelines (USCG 2007), but engineers within Coast Guard ranks must also adhere to the ethical standards that their profession imposes. Thus they are obliged to take action to ensure safety and effectiveness of navigation projects and candidly and fully inform management when activities violate the engineering code.

CHAPTER 10

NOAA CONTRIBUTIONS TO WATERWAYS[1]

10.1 GENERAL

The National Oceanic and Atmospheric Administration (NOAA) supports navigable waterways through its operating branches, most notably the National Ocean Service, the National Weather Service, and the National Environmental, Satellite, Data, and Information Service, and also through the National Marine Fisheries Service.

Marine navigation tools are necessary to ensure safe and efficient marine transportation and commerce, offshore engineering projects, naval operations, and recreational activities. The Office of Coast Survey (OCS), which is part of NOAA's National Ocean Service (NOS), is responsible for providing tools such as nautical charts and hydrographic surveys.

10.2 NAUTICAL CHARTS

OCS remains the primary agency responsible for constructing and maintaining the nation's coastal nautical charts. The Corps of Engineers produces "navigational maps" for some inland rivers, primarily the Mississippi, Ohio, Tennessee, Columbia, and their tributaries. Nautical charts contain information about the nature and shape of the coast, the depth of the water, and general character and configuration of the sea bottom, locations of dangers to navigation, the rise and fall of the tides, locations of navigational aids, and characteristics of the Earth's magnetism. The

[1]Material in this chapter has been extracted from PORTS: Physical Oceanographic Real-Time System (NOAA 2007a) and ASCE Manual No. 107 (McCartney et al. 2005).

charts are compiled by using a fleet of hydrographic vessels that operate in the continental United States, Alaska, and Hawaii.

NOS collects marine hydrographic data (depth soundings) to construct and maintain more than 1,000 nautical charts. In addition, NOS makes available an historical map and chart collection—more than 20,000 maps and charts dating from the late 1700s. The collection includes nautical charts, hydrographic surveys, topographic surveys, geodetic surveys, city plans, and Civil War battle maps.

Deep-water (ocean) hydrographic surveys are made from ships as shown in Fig. 10-1, whereas shallow protected waters (bays, inlets, tidal rivers) are surveyed using launches as shown in Fig. 10-2.

10.3 TIDES AND CURRENTS

NOS has been monitoring sea level variations for many years. For some U.S. locations, sea level records exist for more than 100 years. Water level data are used for a variety of practical purposes, including hydrography, nautical charting, maritime navigation, coastal engineering, and tsunami and storm surge warnings. Mariners use the information to time their approach to and exit from U.S. ports. Long-term applications include marine boundary determinations, tidal predictions, monitoring sea level trends, oceanographic research, and climate research. Bridge, breakwater, and deep-water channel construction also are affected by tidal and current changes.

FIGURE 10-1. NOAA Ship Ronald H. Brown *at Newport, Oregon. Courtesy of NOAA.*

FIGURE 10-2. Hydrographic Survey Launch off NOAA Ship Pierce, *Penobscot Bay, Maine. Courtesy of NOAA.*

Within NOS, the Center for Operational Oceanographic Products and Services (CO-OPS) is primarily responsible for predicting and measuring water levels and currents, and disseminating this information. CO-OPS collects, analyzes, and distributes such data to maintain safe maritime navigation and waterborne commerce. This real-time information is provided to shipmasters and pilots to help avoid groundings and collisions. The information provided includes water levels, currents, and other oceanographic and meteorological data from bays and harbors via telephone voice response and the Internet.

CO-OPS also manages the nation's National Water Level Observation Network (NWLON). NWLON provides basic tidal information to determine U.S. coastal marine boundaries and to create nautical charts. It also supports climate monitoring activities, tsunami and storm surge warning systems, coastal processes, and tectonic research. It consists of 175 continuously operating water level measurement stations along the U.S. coasts and in the Great Lake regions. Many of these stations have been operational and transmitting data for at least 19 years.

10.4 CURRENTS

Water currents are more difficult to measure than water levels. In the past, observations of currents were made for only a few days at a time at any particular location. More recently, however, continuous current observations have been made at several locations along the nation's coasts; these observation stations are subject to corrosion, marine fouling, and other damage, and are expensive to maintain.

Current measurements are usually taken by acoustic Doppler meters, which emit an acoustic signal from either a boat or bottom-mounted transmitter. The signal is reflected from sediment or other particles transported by the flow and recorded. The reflected signal is analyzed to detect the Doppler shift in frequency, yielding a measure of flow velocity in three dimensions.

10.5 GLOBAL POSITIONING

The Global Positioning System (GPS) includes a constellation of 24 satellites, launched and operated by the U.S. Air Force, which transmits radio signals. When used according to standardized procedures, GPS receivers can determine positional coordinates to centimeter-level accuracy, on the horizontal, anywhere on the surface of the Earth. The first GPS satellite was launched in 1978 and the system was declared fully operational for civilian applications in December, 1993.

Augmenting this space-based system is a network of Continuously Operating GPS Reference Stations (CORS), which serve as the foundation for the National Spatial Reference System (NSRS). NSRS is a coordinate system that defines position (latitude and longitude), elevation, distance and direction between points, strength of gravitational pull, and the way in which these values change over time. This information is essential for ensuring the reliability of transportation and communication systems, boundary and property surveys, land record systems, mapping and charting, and many scientific and engineering applications. NSRS provides the positional integrity that allows use of GPS for many modern positioning applications.

The National Geodetic Survey (NGS), part of NOAA's National Ocean Service, coordinates a network of more than 400 CORS stations which receive GPS radio signals 24 hours a day, 7 days a week. The GPS data collected at these stations allow GPS users to determine more accurate positions through computation after the data are collected.

10.6 COAST PILOT

The U.S. Coast Pilot consists of a series of nautical books that cover information important to navigators of coastal and intercoastal waters and the Great Lakes. Issued in nine volumes, these books contain supplemental information that is difficult to portray on a nautical chart.

Topics in the Coast Pilot include channel descriptions, anchorages, bridge and cable clearances, currents, tide and water levels, prominent features, pilotage, towage, weather, ice conditions, wharf descriptions, dangers, routes, traffic separation schemes, small-craft facilities, and federal regulation applicable to navigation. Coast Pilot publications are available through NOAA-authorized network nautical agents.

10.7 PORTS

The Physical Oceanographic Real-Time System (PORTS) is an information acquisition and dissemination technology developed by the NOS in cooperation with a number of ports throughout the United States. The first permanent, fully integrated, operational PORTS was deployed in Tampa Bay during 1990 and 1991. The system is managed, operated, and maintained under a cooperative agreement with NOS. PORTS includes the integrations of real-time currents, water levels, winds, and water temperatures at multiple locations with a data dissemination system that includes telephone voice response as well as modem dial-up and dedicated modem displays. PORTS consists of Acoustic Doppler Current Profilers (ADCPs) with water temperature sensors, a "nowcast" of currents at other locations, water level gauges with anemometers, packet radio transmission equipment, a data acquisition system, and an information dissemination system (IDS).

The traditional prediction tables that are updated annually by NOAA provide information about the astronomical tides, currents, river flows, and other meteorological forces. Real-time measurements, enriched by nowcasts, were identified as critical requirements for safe navigation, leading to the creation of the PORTS system.

PORTS is a public information system that provides real-time information to the general public and provides essential information for safe and cost-effective navigation, search-and-rescue, hazardous material and oil spill prevention and response, and scientific research. PORTS also provides NOAA's Global Ocean Observing System with coastal ocean measurement and dissemination components. All data are continuously archived and are available to the public. PORTS data are broadcast over NOAA Weather Radio hourly by the National Weather Service and are available on a priority basis for trajectory modeling in support of the U.S. Coast Guard. PORTS systems were operational at the following locations in 2008:

- Cherry Point
- Chesapeake Bay
- Lower Columbia River

- Delaware River and Bay
- Houston/Galveston
- Los Angeles/Long Beach
- Mobile Bay
- Narragansett Bay
- New Haven
- New York/New Jersey Harbor
- Pascagoula
- Port of Anchorage
- Sabine-Neches
- San Francisco Bay
- Soo Locks
- Tacoma
- Tampa Bay

10.8 MARINE AND COASTAL WEATHER SERVICES

The NOAA Weather Radio network provides voice broadcasts of local and coastal marine forecasts on a continuous cycle. The forecasts are produced by local National Weather Service forecast offices. Coastal stations also broadcast predicted tides and real-time observations from buoys and coastal meteorological sensors operated by NOAA's National Data Buoy Center. Recorded voice broadcasts have been largely supplemented by a computer-synthesized voice. The NOAA Weather Radio network provides near-continuous coverage of the coastal United States, the Great Lakes, and Hawaii, and can extend much farther in certain areas.

10.9 ETHICAL CONSIDERATIONS

Ethical situations in NOAA activities are substantially similar to those of any engineer working in navigation project operations. Many issues will center on adequate funding and staffing to conduct the agency's missions, as, for example, in maintaining nautical charts. Members of the NOAA Commissioned Officer Corps are governed by an internal set of ethical guidelines (NOAA 2007b) but engineers within NOAA ranks must also adhere to the ethical standards that their profession imposes. Thus they are obliged to take action to ensure safety and effectiveness of navigation projects and candidly and fully inform management when activities violate the engineering code.

CHAPTER 11

TOOLS TO ENSURE SAFE DESIGN AND OPERATION

11.1 GENERAL

In the planning, design, and operations stages of navigation projects, the engineer has a wide variety of tools available to aid in ensuring that the project provides for the ". . . safety, health, and welfare of the public" and sustainability and meets the other requisite criteria. During the planning stage the feasibility of the project is evaluated along with design alternatives to best suit the project's goals and objectives, project site limitations, applicable statutes and ordinances, environmental sustainability, and funding. It is the responsibility of the engineer to select the best tools available that will provide the necessary results within the scheduled timeframe. Ethically speaking, the engineer needs to make those selections based on best professional judgment, availability of specific tools and experienced staff and adequate facilities, and funding and time constraints. The more than can be learned of potential project performance in the planning and design stages, the greater the potential for project success upon completion.

The tools available to the engineer are almost limitless; however, the process of using those tools can also potentially be an almost endless process, so selection must be undertaken prudently and with forethought for the final project and inevitable constraints. The available tools include:

- Standard and accepted hydraulic, sediment, and environmental computations.
- Physical (scale) model studies, in both the near- and far-field.
- Numerical model studies, in both the near- and far-field.
- Vessel simulation studies.
- Field (prototype) evaluations.

- Other useful and helpful techniques available such as spreadsheets, screening tools, and similar procedures.

Case studies are also an important tool to evaluate and apply lessons learned (e.g., the post-Hurricane Katrina assessment mentioned in Chapter 4).

Probably the biggest key to success in using these available tools is tied to the earlier discussion on ethics that the engineer maintains competence relative to education and experience with the specific tools used during the planning and design stages. Although it is not necessarily a requirement that the design engineer have competence in all of the tools used on that project, the engineer should have an understanding of such and should also ensure that the personnel using the tools are in fact competent to do so. Likewise, to maintain high ethical standards the engineer should ensure that input data obtained from the prototype for computations or model studies to be undertaken should be as accurate and pertinent as possible. If the engineer determines that gaps or omissions are present in the prototype input data, that situation should be noted to the client and requests made for additional data.

11.2 STANDARD AND ACCEPTED COMPUTATIONS

These are the equations and formulas most often initially learned in an engineer's undergraduate studies and enhanced through graduate studies and/or experience working on navigation projects. The principles and guidance pertaining to such computations are widely accepted and provide a firm basis for, at a minimum, the initial design considerations. Computations using the Equation of Continuity (e.g., $Q = A * V$), the Equation of Energy Conservation (e.g., the Bernoulli equation), the Froude number (to determine flow conditions of subcritical, critical, or supercritical), and similar equations are indicative of the tools available to the engineer. For example, Parchure et al. (2001) used simple equations for sediment erosion to make desktop calculations of sediment resuspension by passing tows on the Upper Mississippi River as part of studies to evaluate environmental effects of waterway traffic.

11.3 PHYSICAL MODEL STUDIES

Physical model studies have historically been used to evaluate navigability and/or sedimentation issues relative to navigation projects. Such models have a long history of providing exceptional data and information to the engineer in the design and planning of navigation projects, as well as providing useful data and information addressing operational problems on completed projects such as shoaling in lock approaches or addition of a hydropower plant at an existing lock and dam. Models that address near-field issues are typically referred to as "section," "spillway," or "stilling basin" models and consider local concerns such as stone protection requirements, cavitation, hydraulic loads on structure appurtenances, and similar properties. Figure 11-1 presents a photograph of a typical section-type model of the Bluestone Lake Dam on the New River in West Virginia.

Models that address far-field issues are often referred to as "general" models and usually reproduce several miles (or kilometers) of the river under study, with the navigation project located near the center (upstream to downstream) in that model. Figure 11-2 is a photograph of the general, movable-bed model of Olmsted Locks and Dam, Ohio River being used to investigate navigation and sedimentation conditions during and after construction. ASCE Manual No. 97, Hydraulic Modeling: Concepts and Practice (ASCE 2000) provides an excellent resource on modeling, as does Franco (1978) concerning details to be addressed during physical, movable-bed model studies. Relative to attributes useful in the design and planning of navigation projects, ASCE Manual No. 94, Inland Navigation: Locks, Dams, and Channels (McCartney et al. 1995) for shallow-draft projects and the Corps of Engineers EM 1110-2-1613, Hydraulic Design of Deep-Draft Navigation Projects (USACE 2006a) provide exceptional information and guidance that the engineer will find useful in addressing project issues.

FIGURE 11-1. 1:36 Scale Section Model of Bluestone Lake Dam, New River, West Virginia. Courtesy of U.S. Army Corps of Engineers.

FIGURE 11-2. General, Movable-Bed Model of Olmsted Locks and Dam, Ohio River. Courtesy of U.S. Army Corps of Engineers.

11.4 NUMERICAL MODEL STUDIES

Over the past few decades the development and use of numerical hydraulic models has blossomed and many of the issues that previously were only addressed using hand calculations or physical models can now be addressed using numerical models. Numerical models apply numerical techniques such as iteration to solve simplified versions of complex equations of motion, such as the Reynolds-averaged Navier-Stokes equations for conservation of fluid momentum. They produce detailed pictures of flow in rivers, estuaries, and coastal waters under past or future conditions, which are then used in design and evaluation of navigation projects. For example, McAnally and Pritchard (1997) examined Corps of Engineers numerical studies of an earthen sill placed in the Mississippi River to limit salt water intrusion, and found that the models accurately predicted the interaction of the sill and river flows. Typical models used in navigation studies include Corps models such as HEC-RAS, TABS-MD, and ADCIRC, which are supported by the Watershed Modeling System (WMS) and Surface Water Modeling System (SMS) (ERDC 2007).

There is a very wide range of numerical models useful in the design, planning, and operation of navigation projects addressing the hydraulics in one, two, and three dimensions. Certain numerical models also address sedimentation issues and have the additional advantage over physical, movable-bed models in that suspended sediment transport can be addressed numerically even though it was virtually impossible (or at best extremely difficult) to do physically. Numerical models are also useful in addressing selective withdrawal issues from reservoirs, as well as hydrothermal concerns from reservoirs and navigation dam pools. ASCE Manual No. 97, Hydraulic Modeling: Concepts and Practice (ASCE 2000, p. 23) has a section addressing numerical modeling. Figure 11-3 is a portion of a general research, idealized inlet study being conducted by the Corps of Engineers at the Engineer Research and Development Center (ERDC) located at the Waterways Experiment Station (WES) in Vicksburg, Mississippi. The jetties shown could be developed and studied in either a physical or numerical model.

FIGURE 11-3. Jetties under Study on Idealized Inlet Research Study at WES. Courtesy of U.S. Army Corps of Engineers.

11.5 VESSEL SIMULATION STUDIES

Almost simultaneously with the development and use of numerical hydraulic models, vessel simulation capabilities also were being developed and applied to navigation projects. The ERDC at WES set up the first Ship/Tow Simulator for use in the design of navigation projects for the Corps. The WES Ship/Tow Simulator can be used to simulate ports or harbors, inland waterways, and other maritime environments. The models used on the simulator accurately produce flow currents, wind and wave conditions, shallow-water effects and bank forces (when applicable), ship handling, ship-to-ship interaction (in a meeting and passing or overtaking and passing situation), fender forces, anchor forces, and tug assistance. The two simulators at WES can be operated independently or integrated to represent one virtual world. In an integrated simulation, the pilots controlling the two simulators interact with each other via radio and through the visual scene. The normal testing procedure is for WES researchers to set up and adjust the simulator for the project under study, have actual pilots come to WES to verify the simulator study by "navigating" the reach, have the WES researchers develop corrective plans, and have another set of pilots come to WES to evaluate improvement schemes. Figure 11-4 is a photograph of a New York pilot at the WES Ship/Tow Simulator pilot

FIGURE 11-4. A Ship Pilot "Navigating" a Scene on the ERDC/WES Ship/Tow Simulator. Courtesy of U.S. Army Corps of Engineers.

control panel with the visual display of the project under study in the background.

11.6 FIELD EVALUATION STUDIES

Field investigations involve measuring physical or biological conditions at a site and analyzing the measurements. They are used to define conditions at a site, to provide boundary conditions for a numerical model, and to diagnose problems in navigability. In certain instances it is advantageous to use the prototype to study a problem that develops at a completed navigation project. One excellent method used at WES is time-lapse video recording. This method has proved very successful in evaluating navigation into navigation locks on the Ohio and Mississippi Rivers. It was also used to evaluate navigation conditions and towboat paths on the Gulf Intracoastal Waterway and at Greenville Bridge on the Mississippi River. This particular study method provides real-world conditions and has been very helpful in documenting and solving problems on navigation projects.

11.7 TIERED ANALYSIS

The rise of easy-to-use numerical models has led to a similar rise in complaints that some users treat them as "black boxes" or as if the model were one of those magic balls where the answer to any question mysteriously appears in a tiny window. In truth, the phenomenon of mindlessly accepting answers from any tool, be it a slide rule or a physical model, has always been with us; however, numerical models have evolved to the point that we can generate many more incorrect answers than ever before. A solution to the black box problem exists in the form of an old solution technique—the tiered analysis, sometimes called the phased approach.

In its basic form, tiered analysis consists of at least three steps:

- Get ballpark results.
- Get approximate results.
- Rigorous results.

At each step, ask, "Is this result reasonable? Does it agree with other steps?" Be able to explain the differences among the results.

For example, if we needed to predict water levels in a waterway after deepening of the navigation channel, we might get ballpark results using Manning's equation for steady, uniform flow in a prismatic channel. Approximate results could be obtained by setting up and running a schematic numerical model with simplified unsteady inflow and tailwater elevation. In the final step, rigorous results, we might set up and run a complex unsteady, non-uniform flow numerical model with high resolution of geometry and realistic time-varying inflows and tailwater elevations.

Although it might seem that solving the problem three times would take longer, the authors' experience has been that a useful end product occurs sooner because time-consuming problems are discovered early, when they can be corrected more easily. Furthermore, by getting a solution three times, the engineer gains a better understanding of the processes and the project and is better able to exercise good engineering judgment—the antithesis of the black box syndrome. The benefits are fewer major mistakes, faster results, and a better product.

11.8 RISK ASSESSMENT AND UNCERTAINTY

As discussed earlier in this Manual, risk assessment and uncertainty relative to navigation projects need to be addressed during the design process. Ethically, the engineer should be aware of the impacts if the project does not perform as planned and designed. The extent to which a project will not reach it goals is a direct function of the risks associated with the design and the unknowns, such as changes in conditions over time. A schematic risk analysis chart comparing the probability of failure versus the consequences of failure is shown in Fig. 11-5. Tuholski et al. (2002) developed a risk assessment methodology for construction of Braddock Dam on the Monongahela River near Pittsburgh; Table 11-1 shows their simple example of how the risk assessment might be constructed for a typical project. Table 11-1 can be used to construct a matrix of risk like that shown in Fig. 11-5.

It is important for the engineer to evaluate the consequences of inadequate project performance relative to the probability of failure, and model studies provide the engineer with excellent tools to make such evaluations. The engineer should consider ASCE Manual No. 107, Ship Channel Design and Operation (McCartney et al. 1995) for information relative to traffic flow modeling which would be useful in evaluating traffic flow as

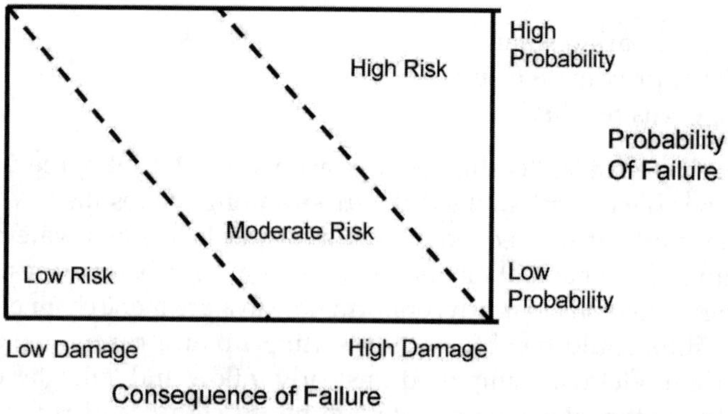

FIGURE 11-5. Schematic Risk Assessment Diagram.

TABLE 11-1. Examples of Occurrence (left) and Consequence (right) Classifications to Be Adopted in a Qualitative Risk Analysis

Occurrence Class	Frequency of Occurrence	Consequence Class	Consequence
Low	"<0.01 per year" or "Less than 1 occurrence per 100 projects"	Low	"<\$100,000" or "Less than 5 injuries"
Medium	"0.01–1.0 per year" or "More than 1 occurrence per 100 projects, but less than 1 occurrence per project"	Medium	"\$100,000–\$1 million" or "More than 5 injuries but fewer than 2 fatalities"
High	">1.0 per year" or "More than once per project"	High	">\$1 million" or "More than 2 fatalities"

Source: Tuholski et al. (2002). Courtesy of U.S. Army Corps of Engineers.

well as the potential for encounters (accidents). This will help describe the consequences of failure as well as give a feel for the probability of failure. Also relative to modeling, ASCE Manual No. 97, Hydraulic Modeling: Concepts and Practice (ASCE 2000, p. 326) addresses uncertainty analysis and also provides excellent references for in-depth analysis and insights.

11.9 ETHICAL CONSIDERATIONS

The tools that are available to an engineer to address navigation projects are vast and cover a wide range for necessary studies. For competence and maintaining the integrity necessary to ensure the "safety, health, and welfare of the public," studying the design, planning, and operation of proposed or existing navigation projects is paramount in the ethical engineering approach. Such studies provide the benefits of determining the strong points of a project and, often just as helpful, provide the engineer with the knowledge that some alternatives do not produce desirable results. Proper use of these tools will aid the engineer in having an excellent view on the potential performance of navigation projects, which is ethically what ASCE, the engineering community, and the public expects from observance of Canons 1, 4, and 7 in the ASCE Code of Ethics.

CHAPTER 12
CONCLUSIONS

Engineering criteria and practices for design, operation, and management of navigation projects are founded on and interwoven with engineering ethics.

Navigation projects provide for waterborne transport of people and goods by ships, barges, ferries, and other vessels. They consist of ports, harbors, channels, locks, and related facilities and they constitute vital links in the U.S. Marine Transportation System.

Navigation projects are engineered—designed, constructed, operated, and maintained in accordance with engineering criteria defined in laws, regulations, codes, guidance, and good practice. The common foundation that supports these criteria must be engineering ethics.

Ethical behavior consists of doing the right thing. For engineers, that behavior is guided by professional codes of ethics, licensing laws, organizational standards of conduct, and oaths of office. The ASCE Code of Ethics (Appendix) provides the guidelines and standards members are expected to observe. Of particular note to navigation projects are these Canons of the Code:

- Canon 1: Engineers shall hold **paramount** (emphasis added) the safety, health, and welfare of the public and shall strive to comply with the principles of sustainable development in the performance of their professional duties.
- Canon 2: Engineers shall perform services only in areas of their competence.
- Canon 4: Engineers shall act in professional matters for each employer or client as faithful agents or trustees, and shall avoid conflicts of interest.
- Canon 6: Engineers shall act in such a manner as to uphold and enhance the honor, integrity, and dignity of the engineering profession and shall act with zero tolerance for bribery, fraud, and corruption.

- Canon 7: Engineers shall continue their professional development . . ., and shall provide professional development for engineers under their supervision.

The Code also serves as a standard by which all engineering practice will be judged.

The practice of navigation engineering involves the planning, design, construction, operation, and maintenance of safe, reliable, efficient, and environmentally sustainable navigable waterways (channels, structures, and support systems) used to move people and goods by waterborne vessels.

Engineering design criteria for navigation projects further refine the call for ethical behavior stated under Canon 1. Engineering criteria can be expressed in terms of general principles such as safety, effectiveness, and efficiency, and in more specific terms such as a 1 in 100 probability of accidental property damage. Specific engineering criteria are found in design manuals produced by the U.S. Army Corps of Engineers, ASCE, and other organizations, and virtually all of those criteria can be traced to a specific ethical requirement.

Sustainable water resource systems are those designed and managed to meet the needs of people living in the future as well as those of us living today. Sustainable development is a goal of Canon 1 of the ASCE Code of Ethics and has been incorporated into many organizations' design principles.

The Corps of Engineers, the U.S. Coast Guard, and NOAA each have management roles and ethical responsibilities in navigation systems. The Corps operates and maintains structures and channels and provides resource management in order for projects to meet their intended purposes. The Coast Guard provides for safe navigation and security. NOAA provides services such as tide and weather forecasts to mariners. All of these activities contribute to the effectiveness of the system while safeguarding public welfare and safety, and must always be performed in accordance with Canon 1 of the ASCE Code of Ethics.

Design tools used to evaluate various channel and hydraulic structure configurations include standard computations, physical and numerical models, and field investigations, and are best employed in a tiered analysis approach. Good engineering practice, including proper use of physical models, numerical models, and field investigations plus mathematical and software tools, is defined by the evolving knowledge and skills of the engineering community, in which engineers are required by ASCE Code Canon 7 to maintain currency.

REFERENCES

American Society of Civil Engineers (ASCE). (1994). Planning and Design Guidelines for Small Craft Harbors, Manuals and Reports on Engineering Practice No. 50, American Society of Civil Engineers, New York.

ASCE. (1998). Sustainability Criteria for Water Resource Systems, American Society of Civil Engineers, Reston, VA.

ASCE. (2000). Hydraulic Modeling: Concepts and Practice, Manuals and Reports on Engineering Practice No. 97, American Society of Civil Engineers, Reston, VA.

ASCE. (2004). "Policy Statement 351: Project Peer Review," <http://www.asce.org/pressroom/news/policy_details.cfm?hdlid=17> (May, 2007).

ASCE. (2005). "Report Card for America's Infrastructure—Navigable Waterways," <http://www.asce.org/reportcard/2005/page.cfm?id=36> (May, 2007).

ASCE. (2006a). Hurricane Katrina: One Year Later: What Must We Do Next? A Statement by the American Society of Civil Engineers Hurricane Katrina External Review Panel, American Society of Civil Engineers, Reston, VA.

ASCE. (2006b). "Code of Ethics," <http://www.asce.org/inside/codeofethics.cfm> (Nov., 2006).

ASCE. (2006c). "American Society of Civil Engineers Policies," <http://www.asce.org/pressroom/news/policy_details.cfm?hdlid=60> (Nov., 2006).

ASCE. (2008). Ethics: Guidelines for Conduct for Civil Engineers, American Society of Civil Engineers, Reston, VA.

Building Owners and Managers Institute International (BOMI). (2007). "International Code of Professional Ethics and Conduct" <http://www.bomi-edu.org/EduOfferingscontent.aspx?id=468> (May, 2007).

Calhoun, C. C., Jr. (2007). "Ethics: Rocket Science or Child's Play?" *Leadership and Management in Engineering* 7(1), 24–27, ASCE.

Casavant, K. (2000). Inland Waterborne Transport—An Industry Under Siege, U.S. Department of Agriculture, Agricultural Marketing Service, Washington, D.C.

Center for Study of Ethics in the Professions. (2007). "Codes of Ethics," <http://ethics.iit.edu/codes/> (May, 2007).

Covey, S. R. (1998). "Four Roles of Powerful and Empowering Leadership." Notes, Third Worldwide Lessons in Leadership Series, Lessons in Leadership, Lexington, KY.

d'Angremond, K., and Roode, F. C. (2001). Breakwaters and Closure Dams, Delft University Press, The Netherlands.

Drucker, P. F. (1999). Management Challenges for the 21st Century, HarperBusiness, New York.

ERDC. (2007). "Coastal and Hydraulics Laboratory Software," <http://chl.erdc.usace.army.mil/software> (May, 2008).

Franco, J. J. (1978). Guidelines for the Design, Adjustment and Operation of Models for the Study of River Sedimentation Problems, Instruction Report H-78-1, U.S. Army Corps of Engineers Waterways Experiment Station, Vicksburg, MS.

Garrett, M. F. (2006). "What's the Use?" *Leadership and Management in Engineering* 6(4), 173–174, ASCE.

Goldberg, D. E. (2006). The Entrepreneurial Engineer, John Wiley & Sons, Hoboken, NJ.

Huntsman, J. M. (2005). Winners Never Cheat, Wharton School Publishing, Upper Saddle River, NJ.

Jefts, A. R. (2003). "Houston-Galveston Navigation Channels, Texas Project 50-Year Plan Development and Design, Dredging '02: Key Technologies for Global Prosperity," Proceedings of the Third Specialty Conference on Dredging and Dredged Material Disposal, May 5–8, 2002, Orlando, FL, Stephen Garbaciak, Jr., ed., American Society of Civil Engineers, Reston, VA

Maxwell, J. C. (2003). There Is No Such Thing as "Business Ethics," Warner, New York.

McAnally, W. H., and Pritchard, D. P. (1997). "Salinity Control in the Mississippi River under Drought Flows." *Journal of Waterway, Port, Coastal, and Ocean Engineering* 123(1), 34–40, ASCE.

McAnally, W. H., Zhang, Y., Allen, A. J., Bowden, R. O., and Tan, A. (2004). Transportation Responses to Increased Latin American Trade, National Center for Intermodal Transportation, Mississippi State University, Starkville, MS.

McCartney, B. L., Ebner, L., Hales, L. Z., and Nelson, E. E. (2005). Ship Channel Design and Operation. Manuals and Reports on Engineering Practice No. 107, American Society of Civil Engineers, Reston, VA.

McCartney, B. L., George, J., Lee, B. K., Lindgren, M., and Neilson, F. (1995). Inland Navigation: Locks, Dams, and Channels, Manuals and Reports on Engineering Practice No. 94, American Society of Civil Engineers, Reston, VA.

McCartney, B. L., Beeman, O., Herrmann, Jr., F. A., and Hopman, R. J. (1993). Report on Ship Channel Design, Manuals and Reports on Engineering Practice No. 80, American Society of Civil Engineers, New York.

National Oceanic and Atmospheric Administration (NOAA). (2007a). "PORTS: Physical Oceanographic Real-Time System—A Tool for Safe Navigation," <http://chartmaker.ncd.noaa.gov/psn/ports.htm> (May, 2007).

NOAA. (2007b). "NOAA Corps Values," <http://www.noaacorps.noaa.gov/about/values.html> (May, 2007).

Parchure, T. M., McAnally, W. H., and Teeter, A. M. (2001) "Desktop Method for Estimating Vessel-Induced Sediment Suspension." *Journal of Hydraulic Engineering* 127(7), 577–587, ASCE.

Pfatteicher, S. K. A. (2003). "Depending on Character: ASCE Shapes Its First Code." *J. Prof. Issues Eng. Educ. Pract.* 1291, 21–31.

International Navigation Association (PIANC). (1997). Joint PIANC-IAPH Report on Approach Channels—A Guide for Design (Volume 2), Marcom Working Group 30, International Navigation Association, Brussels.

PIANC. (2003). Guidelines for Sustainable Inland Waterways and Navigation, Report of Working Group 6 of the Environmental Commission, International Navigation Association, Brussels.

Texas Board of Professional Engineers (Texas). (2006). Texas Engineering Practice Act and Rules Concerning the Practice of Engineering and Professional Engineering Licensure, Texas Board of Professional Engineers, Austin, TX.

Tuholski, N. J., Gluver, H., Cornell, C. A., Gerwick, Jr., B. C., Patev, R. C., and Padula, J. A. (2002). Risk Assessment Procedures for Innovative Navigation Projects, ERDC/ITL TR 02-4, U.S. Army Engineer Research and Development Center, Vicksburg, MS.

Tennessee Valley Authority (TVA). (2006). "TVA Environmental Policies and Principles," <http://www.tva.gov/environment/policy.htm> (Nov., 2006).

U.S. Army Corps of Engineers (USACE). (1980). Layout and Design of Shallow-Draft Waterways, EM 1110-2-1611, U.S. Army Corps of Engineers, Washington, D.C.

USACE. (1981). Deep Draft Navigation Project Design, ER 1110-2-1404, U.S. Army Corps of Engineers, Washington, D.C.

USACE (1984). Hydraulic Design of Small Boat Harbors, EM 1110-2-1615, U.S. Army Corps of Engineers, Washington, D.C.

USACE. (1985). Hydraulic Design of Small Boat Navigation Projects, ER 1110-2-1457, U.S. Army Corps of Engineers, Washington, D.C.

USACE. (1986). Design of Breakwaters and Jetties, EM 1110-2-2904, U.S. Army Corps of Engineers, Washington, D.C. (Rescinded)

USACE. (1987). Beneficial Uses of Dredged Material, EM 1110-2-5026, U.S. Army Corps of Engineers, Washington, D.C.

USACE. (1991). Tidal Hydraulics, EM 1110-2-1607, U.S. Army Corps of Engineers, Washington, D.C.

USACE. (1995a). Planning and Design of Navigation Locks, EM 1110-2-2602, U.S. Army Corps of Engineers, Washington, D.C.

USACE. (1995b). Hydraulic Design of Navigation Locks, EM 1110-2-1604, U.S. Army Corps of Engineers, Washington, D.C.

USACE. (1996a). Flood Control Operations and Maintenance Policies, ER 1130-2-530, U.S. Army Corps of Engineers, Washington, D.C.

USACE. (1996b). Channel Maintenance Management Plan, U.S. Army Corps of Engineers, St. Paul District, St. Paul, MN.

USACE (1996c). Hydraulic Design of Deep Draft Navigation Projects, ER 1110-2-1404, U.S. Army Corps of Engineers, Washington, D.C.

USACE. (1997). Hydraulic Design for Coastal Shore Protection Projects, ER 1110-2-1407, U.S. Army Corps of Engineers, Washington, D.C.

USACE. (1998). Hydraulic Design of Shallow Draft Navigation Projects, ER 1110-2-1458, U.S. Army Corps of Engineers, Washington, D.C.

USACE. (2000). Planning Guidance Notebook, ER 1105-2-100, Chapter 2, "Planning Principles," U.S. Army Corps of Engineers, Washington, D.C.

USACE. (2001). Emergency Employment of Army and Other Resources—Civil Emergency Management Program—Procedures, Engineering Pamphlet 500-1-1, Chapter 3, "Disaster Preparedness," U.S. Army Corps of Engineers, Washington, D.C.

USACE. (2003). Safety—Safety and Health Requirements, EM 385-1-1, U.S. Army Corps of Engineers, Washington, D.C.

USACE. (2004). Safety of Dams—Policy and Procedures, ER 1110-2-1156, Chapter 6, "Periodic Inspection and Continuing Evaluation," U.S. Army Corps of Engineers, Washington, D.C.

USACE. (2005a). "Environmental Operating Principles," <http://www.hq.usace.army.mil/cepa/pubs/apr05/STORY5a.htm> (May, 2008).

USACE. (2005b). UMR-IWW System Navigation Study Newsletter, September, Vol. 11, No. 2, "Projects at a Glance," U.S. Army Corps of Engineers, Rock Island, IL.

USACE. (2006a). Hydraulic Design of Deep-Draft Navigation Projects, EM 1110-2-1613, U.S. Army Corps of Engineers, Washington, D.C.

USACE. (2006b). "U.S. Army Corps of Engineers Releases Its "12 Actions for Change." Press Release No. PA-06-11, August 24, <http://www.hq.usace.army.mil/cepa/releases/actionsforchange.htm> (June, 2007).

USACE. (2006c). "History of the Regulatory Program," <http://www.usace.army.mil/cw/cecwo//reg/reghist.pdf> (Nov., 2006).

USACE. (2006d). "Index of Engineer Manuals," <http://www.usace.army.mil/publications/eng-manuals/> (Nov., 2006).

USACE. (2006e). Environmental Engineering for Deep-Draft Navigation Projects, EM 1110-2-1202, U.S. Army Corps of Engineers, Washington, D.C.

USACE. (2007). Project Fact Sheet, "Backlog of Maintenance—Major Rehabilitation and Major Maintenance—Mississippi River & Illinois Waterway Locks and Dams," September, 2006, U.S. Army Corps of Engineers, Rock Island, IL.

U.S. Coast Guard (USCG). (2002). U.S. Coast Guard: America's Maritime Guardian, Coast Guard Publication 1, Washington D.C.

USCG. (2006). "Waterways Action Plan Web Site," <http//www.uscg.mil/d8/Divs/M/D8mwm.htm> (Nov., 2006).

USCG. (2007). "Ethics/Standards of Conduct," <http://www.uscg.mil/mlclant/ldiv/Ethics_and_standards.htm> (May, 2007).

U.S. Department of the Army (USDOA). (2001). Engineering Pamphlet 500-1-1, Army Inspection Policy, Chapter 3, "Army Inspections," U.S. Department of the Army, Washington, D.C.

U.S. Fish and Wildlife Service (USFWS). (2004).Water Resources Development under the Fish and Wildlife Coordination Act, Chapter IV, "Corps of Engineers Project Planning and Implementation," U.S. Fish and Wildlife Service, Arlington, VA.

Veach, C. M. (2006). "There's No Such Thing as Engineering Ethics." *Leadership and Management in Engineering* 6(3), 97–101, ASCE.

Walesh, S. G. (2000). Engineering Your Future, Second Ed., American Society of Civil Engineers, Reston, VA.

Welch, J. (2003). Jack: Straight from the Gut, Business Plus Publishing, New York.

APPENDIX

ASCE CODE OF ETHICS[1]

FUNDAMENTAL PRINCIPLES[2]

Engineers uphold and advance the integrity, honor and dignity of the engineering profession by:

1. using their knowledge and skill for the enhancement of human welfare and the environment;
2. being honest and impartial and serving with fidelity the public, their employers and clients;
3. striving to increase the competence and prestige of the engineering profession; and
4. supporting the professional and technical societies of their disciplines.

FUNDAMENTAL CANONS

1. Engineers shall hold paramount the safety, health and welfare of the public and shall strive to comply with the principles of sustainable development[3] in the performance of their professional duties.

[1]The Society's Code of Ethics was adopted on September 2, 1914, and was most recently amended on July 23, 2006. Pursuant to the Society's Bylaws, it is the duty of every Society member to report promptly to the Committee on Professional Conduct any observed violation of the Code of Ethics.

[2]In April 1975, the ASCE Board of Direction adopted the fundamental principles of the Code of Ethics of Engineers as accepted by the Accreditation Board for Engineering and Technology, Inc. (ABET).

[3]In November 1996, the ASCE Board of Direction adopted the following definition of Sustainable Development: "Sustainable Development is the challenge of meeting

2. Engineers shall perform services only in areas of their competence.
3. Engineers shall issue public statements only in an objective and truthful manner.
4. Engineers shall act in professional matters for each employer or client as faithful agents or trustees, and shall avoid conflicts of interest.
5. Engineers shall build their professional reputation on the merit of their services and shall not compete unfairly with others.
6. Engineers shall act in such a manner as to uphold and enhance the honor, integrity, and dignity of the engineering profession and shall act with zero tolerance for bribery, fraud, and corruption.
7. Engineers shall continue their professional development throughout their careers, and shall provide opportunities for the professional development of those engineers under their supervision.

GUIDELINES TO PRACTICE UNDER THE FUNDAMENTAL CANONS OF ETHICS

CANON 1. Engineers shall hold paramount the safety, health and welfare of the public and shall strive to comply with the principles of sustainable development in the performance of their professional duties.

a. Engineers shall recognize that the lives, safety, health and welfare of the general public are dependent upon engineering judgments, decisions and practices incorporated into structures, machines, products, processes and devices.
b. Engineers shall approve or seal only those design documents, reviewed or prepared by them, which are determined to be safe for public health and welfare in conformity with accepted engineering standards.
c. Engineers whose professional judgment is overruled under circumstances where the safety, health and welfare of the public are endangered, or the principles of sustainable development ignored, shall inform their clients or employers of the possible consequences.
d. Engineers who have knowledge or reason to believe that another person or firm may be in violation of any of the provisions of Canon 1 shall present such information to the proper authority in writing and shall cooperate with the proper authority in furnishing such further information or assistance as may be required.

---

human needs for natural resources, industrial products, energy, food, transportation, shelter, and effective waste management while conserving and protecting environmental quality and the natural resource base essential for future development."

e. Engineers should seek opportunities to be of constructive service in civic affairs and work for the advancement of the safety, health and well-being of their communities, and the protection of the environment through the practice of sustainable development.
f. Engineers should be committed to improving the environment by adherence to the principles of sustainable development so as to enhance the quality of life of the general public.

CANON 2. Engineers shall perform services only in areas of their competence.

a. Engineers shall undertake to perform engineering assignments only when qualified by education or experience in the technical field of engineering involved.
b. Engineers may accept an assignment requiring education or experience outside of their own fields of competence, provided their services are restricted to those phases of the project in which they are qualified. All other phases of such project shall be performed by qualified associates, consultants, or employees.
c. Engineers shall not affix their signatures or seals to any engineering plan or document dealing with subject matter in which they lack competence by virtue of education or experience or to any such plan or document not reviewed or prepared under their supervisory control.

CANON 3. Engineers shall issue public statements only in an objective and truthful manner.

a. Engineers should endeavor to extend the public knowledge of engineering and sustainable development, and shall not participate in the dissemination of untrue, unfair or exaggerated statements regarding engineering.
b. Engineers shall be objective and truthful in professional reports, statements, or testimony. They shall include all relevant and pertinent information in such reports, statements, or testimony.
c. Engineers, when serving as expert witnesses, shall express an engineering opinion only when it is founded upon adequate knowledge of the facts, upon a background of technical competence, and upon honest conviction.
d. Engineers shall issue no statements, criticisms, or arguments on engineering matters which are inspired or paid for by interested parties, unless they indicate on whose behalf the statements are made.
e. Engineers shall be dignified and modest in explaining their work and merit, and will avoid any act tending to promote their own interests at the expense of the integrity, honor and dignity of the profession.

CANON 4. Engineers shall act in professional matters for each employer or client as faithful agents or trustees, and shall avoid conflicts of interest.

a. Engineers shall avoid all known or potential conflicts of interest with their employers or clients and shall promptly inform their employers or clients of any business association, interests, or circumstances which could influence their judgment or the quality of their services.
b. Engineers shall not accept compensation from more than one party for services on the same project, or for services pertaining to the same project, unless the circumstances are fully disclosed to and agreed to, by all interested parties.
c. Engineers shall not solicit or accept gratuities, directly or indirectly, from contractors, their agents, or other parties dealing with their clients or employers in connection with work for which they are responsible.
d. Engineers in public service as members, advisors, or employees of a governmental body or department shall not participate in considerations or actions with respect to services solicited or provided by them or their organization in private or public engineering practice.
e. Engineers shall advise their employers or clients when, as a result of their studies, they believe a project will not be successful.
f. Engineers shall not use confidential information coming to them in the course of their assignments as a means of making personal profit if such action is adverse to the interests of their clients, employers or the public.
g. Engineers shall not accept professional employment outside of their regular work or interest without the knowledge of their employers.

CANON 5. Engineers shall build their professional reputation on the merit of their services and shall not compete unfairly with others.

a. Engineers shall not give, solicit or receive either directly or indirectly, any political contribution, gratuity, or unlawful consideration in order to secure work, exclusive of securing salaried positions through employment agencies.
b. Engineers should negotiate contracts for professional services fairly and on the basis of demonstrated competence and qualifications for the type of professional service required.
c. Engineers may request, propose or accept professional commissions on a contingent basis only under circumstances in which their professional judgments would not be compromised.
d. Engineers shall not falsify or permit misrepresentation of their academic or professional qualifications or experience.

e. Engineers shall give proper credit for engineering work to those to whom credit is due, and shall recognize the proprietary interests of others. Whenever possible, they shall name the person or persons who may be responsible for designs, inventions, writings or other accomplishments.

f. Engineers may advertise professional services in a way that does not contain misleading language or is in any other manner derogatory to the dignity of the profession. Examples of permissible advertising are as follows:
   - Professional cards in recognized, dignified publications, and listings in rosters or directories published by responsible organizations, provided that the cards or listings are consistent in size and content and are in a section of the publication regularly devoted to such professional cards.
   - Brochures which factually describe experience, facilities, personnel and capacity to render service, providing they are not misleading with respect to the engineer's participation in projects described.
   - Display advertising in recognized dignified business and professional publications, providing it is factual and is not misleading with respect to the engineer's extent of participation in projects described.
   - A statement of the engineers' names or the name of the firm and statement of the type of service posted on projects for which they render services.
   - Preparation or authorization of descriptive articles for the lay or technical press, which are factual and dignified. Such articles shall not imply anything more than direct participation in the project described.
   - Permission by engineers for their names to be used in commercial advertisements, such as may be published by contractors, material suppliers, etc., only by means of a modest, dignified notation acknowledging the engineers' participation in the project described. Such permission shall not include public endorsement of proprietary products.

g. Engineers shall not maliciously or falsely, directly or indirectly, injure the professional reputation, prospects, practice or employment of another engineer or indiscriminately criticize another's work.

h. Engineers shall not use equipment, supplies, laboratory or office facilities of their employers to carry on outside private practice without the consent of their employers.

CANON 6. Engineers shall act in such a manner as to uphold and enhance the honor, integrity, and dignity of the engineering profession and shall act with zero tolerance for bribery, fraud, and corruption.

a. Engineers shall not knowingly engage in business or professional practices of a fraudulent, dishonest or unethical nature.
b. Engineers shall be scrupulously honest in their control and spending of monies, and promote effective use of resources through open, honest and impartial service with fidelity to the public, employers, associates and clients.
c. Engineers shall act with zero tolerance for bribery, fraud, and corruption in all engineering or construction activities in which they are engaged.
d. Engineers should be especially vigilant to maintain appropriate ethical behavior where payments of gratuities or bribes are institutionalized practices.
e. Engineers should strive for transparency in the procurement and execution of projects. Transparency includes disclosure of names, addresses, purposes, and fees or commissions paid for all agents facilitating projects.
f. Engineers should encourage the use of certifications specifying zero tolerance for bribery, fraud, and corruption in all contracts.

CANON 7. Engineers shall continue their professional development throughout their careers, and shall provide opportunities for the professional development of those engineers under their supervision.

a. Engineers should keep current in their specialty fields by engaging in professional practice, participating in continuing education courses, reading in the technical literature, and attending professional meetings and seminars.
b. Engineers should encourage their engineering employees to become registered at the earliest possible date.
c. Engineers should encourage engineering employees to attend and present papers at professional and technical society meetings.
d. Engineers shall uphold the principle of mutually satisfying relationships between employers and employees with respect to terms of employment including professional grade descriptions, salary ranges, and fringe benefits.

INDEX